Graphene-based Membranes for Mass Transport Applications

Nanoscience & Nanotechnology Series

Editor-in-chief:
Paul O'Brien CBE FREng FRS, *University of Manchester, UK*

Series editors:
Xiaogang Liu, *National University of Singapore, Singapore*
Ralph Nuzzo, *University of Illinois at Urbana-Champaign, USA*
Joao Rocha, *University of Aveiro, Portugal*

Titles in the series:
1: Nanotubes and Nanowires
2: Fullerenes: Principles and Applications
3: Nanocharacterisation
4: Atom Resolved Surface Reactions: Nanocatalysis
5: Biomimetic Nanoceramics in Clinical Use: From Materials to Applications
6: Nanofluidics: Nanoscience and Nanotechnology
7: Bionanodesign: Following Nature's Touch
8: Nano-Society: Pushing the Boundaries of Technology
9: Polymer-based Nanostructures: Medical Applications
10: Metallic and Molecular Interactions in Nanometer Layers, Pores and Particles: New Findings at the Yoctolitre Level
11: Nanocasting: A Versatile Strategy for Creating Nanostructured Porous Materials
12: Titanate and Titania Nanotubes: Synthesis, Properties and Applications
13: Raman Spectroscopy, Fullerenes and Nanotechnology
14: Nanotechnologies in Food
15: Unravelling Single Cell Genomics: Micro and Nanotools
16: Polymer Nanocomposites by Emulsion and Suspension
17: Phage Nanobiotechnology
18: Nanotubes and Nanowires, 2nd Edition
19: Nanostructured Catalysts: Transition Metal Oxides
20: Fullerenes: Principles and Applications, 2nd Edition
21: Biological Interactions with Surface Charge Biomaterials
22: Nanoporous Gold: From an Ancient Technology to a High-Tech Material
23: Nanoparticles in Anti-Microbial Materials: Use and Characterisation
24: Manipulation of Nanoscale Materials: An Introduction to Nanoarchitectonics
25: Towards Efficient Designing of Safe Nanomaterials: Innovative Merge of Computational Approaches and Experimental Techniques
26: Polymer–Graphene Nanocomposites
27: Carbon Nanotube-Polymer Composites
28: Nanoscience for the Conservation of Works of Art
29: Polymer Nanofibers: Building Blocks for Nanotechnology

30: Artificial Cilia
31: Nanodiamond
32: Nanofabrication and its Application in Renewable Energy
33: Semiconductor Quantum Dots: Organometallic and Inorganic Synthesis
34: Soft Nanoparticles for Biomedical Applications
35: Hierarchical Nanostructures for Energy Devices
36: Microfluidics for Medical Applications
37: Nanocharacterisation, 2nd Edition
38: Thermometry at the Nanoscale: Techniques and Selected Applications
39: Nanoceramics in Clinical Use: From Materials to Applications, 2nd Edition
40: Near-infrared Nanomaterials: Preparation, Bioimaging and Therapy Applications
41: Nanofluidics, 2nd Edition
42: Nanotechnologies in Food, 2nd Edition
43: ZnO Nanostructures: Fabrication and Applications
44: Diatom Nanotechnology: Progress and Emerging Applications
45: Nanostructured Materials for Type III Photovoltaics
46: Chemically Derived Graphene: Functionalization, Properties and Applications
47: Graphene-based Membranes for Mass Transport Applications

How to obtain future titles on publication:
A standing order plan is available for this series. A standing order will bring delivery of each new volume immediately on publication.

For further information please contact:
Book Sales Department, Royal Society of Chemistry, Thomas Graham House, Science Park, Milton Road, Cambridge, CB4 0WF, UK
Telephone: +44 (0)1223 420066, Fax: +44 (0)1223 420247
Email: booksales@rsc.org
Visit our website at www.rsc.org/books

Graphene-based Membranes for Mass Transport Applications

Edited by

Pengzhan Sun
University of Manchester, UK
Email: pengzhan.sun@manchester.ac.uk

and

Hongwei Zhu
Tsinghua University, China
Email: hongweizhu@tsinghua.edu.cn

ROYAL SOCIETY OF CHEMISTRY

Nanoscience & Nanotechnology Series No. 47

Print ISBN: 978-1-78262-939-9
PDF ISBN: 978-1-78801-301-7
EPUB ISBN: 978-1-78801-531-8
Print ISSN: 1757-7136
Electonic ISSN: 1757-7144

A catalogue record for this book is available from the British Library

The Royal Society of Chemistry is a charity, registered in England and Wales, Number 207890, and a company incorporated in England by Royal Charter (Registered No. RC000524), registered office: Burlington House, Piccadilly, London W1J 0BA, UK, Telephone: +44 (0) 207 4378 6556.

For further information see our web site at www.rsc.org

Printed in the United Kingdom by CPI Group (UK) Ltd, Croydon, CR0 4YY, UK

Preface

The preparation of graphene-based membrane materials and their applications in mass transport have stimulated intensive research interests and achieved great progress in recent years. However, there lacks a general book focusing on this specific research area. Therefore, in this book, we aim to make a systematic introduction on the state-of-the-art membrane-based filtration and separation technologies, the structure, composition and general properties of graphene and its derivatives (*e.g.* nanoporous graphene and graphene oxide), and especially the selective mass transport properties of graphene-based membranes. For this purpose, we invited several experts who have made substantial achievements in this field to contribute ten chapters in combination with their specific research backgrounds.

The layout of this book is in two main parts: the "understanding" of unique mass transport properties and mechanisms of graphene-based membranes, and their "applications" in various filtration and separation technologies. The first two chapters briefly introduce the state-of-the-art membrane-based separation and filtration processes, as well as the design, fabrication, and characterization of graphene-based membranes. The next three chapters overview the unique mass transport properties and mechanisms of atomically thin graphene-based membranes, graphene oxide lamellar membranes and composite membranes containing graphene oxide sheets. In the final chapters, a few specialized applications of graphene and other two-dimensional material based membranes are covered, *i.e.* barriers, filtration and separation, as well as water desalination.

This book provides a representative example to explain the structure–property–application relationship in materials science research using graphene in membrane-based mass transport as an example. It can serve as an introduction and reference to physicists, chemists, material scientists and

Nanoscience & Nanotechnology Series No. 47
Graphene-based Membranes for Mass Transport Applications
Edited by Pengzhan Sun and Hongwei Zhu

Published by the Royal Society of Chemistry, www.rsc.org

students who are entering or already working in the field of graphene-based membrane materials.

We would like to thank all the authors for contributing their time and expertise in preparing the book chapters. We would like to express our sincere appreciation to Connor Sheppard and Leanne Marle, from the Royal Society of Chemistry, for their help in driving the book's development and eventual publication.

Pengzhan Sun, The University of Manchester, UK
Hongwei Zhu, Tsinghua University, China

Contents

Nanoscience & Nanotechnology Series No. 47
Graphene-based Membranes for Mass Transport Applications
Edited by Pengzhan Sun and Hongwei Zhu

Published by the Royal Society of Chemistry, www.rsc.org

CHAPTER 1

Current State-of-the-art Membrane Based Filtration and Separation Technologies

YANG SU

National Graphene Institute, School of Chemical Engineering and Analytical Science, University of Manchester, Manchester M13 9PL, UK
Email: yang.su@manchester.ac.uk

1.1 Introduction

Chemistry-related manufacturing industries are extensively involved in separation – a process that sorts a mixture into two or more distinct components.[1,2] Such a mixture could be: (1) two or more types of liquids, or gases; (2) liquids in a gas stream; (3) a suspension, that is, one or more solids dispersed in certain liquids; (4) a solution that contains ions and their solvents. In practise, a separation process achieves isolation, purification, or analysis of substances by exploiting the differences of the components, physical properties such as size, mass, and density and chemical properties, and it covers a wide variety of technologies, for example, centrifugation, chromatography, crystallization, distillation, ion exchange, membrane based separation *etc.* Among all the existing technologies, membrane based separation provides efficient, molecular level separation for a large amount of substances, and additionally, it is performed isothermally at ambient temperature, which could lower energy consumption, emissions and pollution compared to other thermal based separation processes, for example, distillation, which accounts for 10–15% global energy consumption (Figure 1.1).[1]

Nanoscience & Nanotechnology Series No. 47
Graphene-based Membranes for Mass Transport Applications
Edited by Pengzhan Sun and Hongwei Zhu

Published by the Royal Society of Chemistry, www.rsc.org

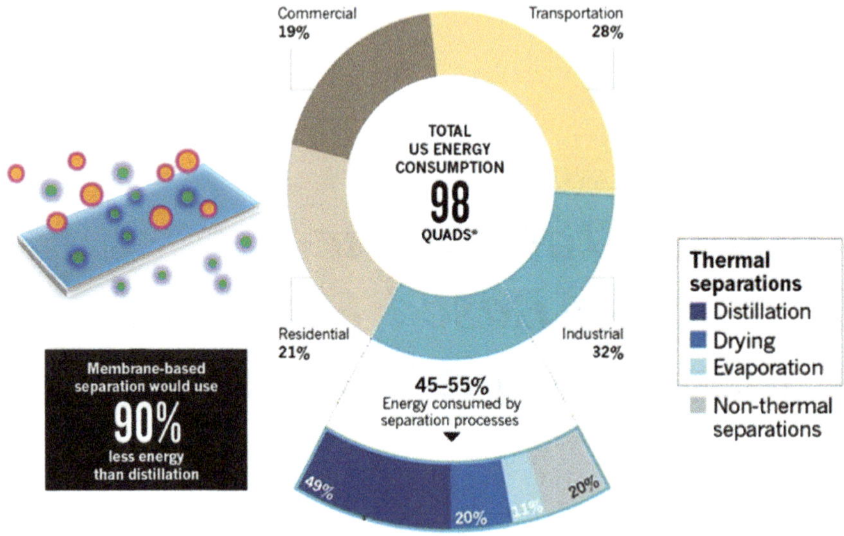

Figure 1.1 Chemical separations account for about half of the United States' industrial energy use. Membrane based separation would largely reduce the energy cost in separation.*a quad is 3×10^{11} kilowatt-hours. Reproduced from ref. 1 with permission from Springer Nature, Copyright 2016.

Therefore, membrane based separation has gained an important role in related industries.

A membrane, which, in essence, provides a permiselective interface that allows or favours one component to permeate, but hinders permeation of others, is a key component in membrane based separation processes. The scientific study on a permiselective membrane could be traced back to 1748, when Abbe Nollet discovered osmosis through a diaphragm from pig's bladder.[3] However, it was not until world war II that the first significant development in membrane application was made, that is, to use synthetic microporous collodion membranes to filter the bacteria in water and evaluate water contamination. Later on, in the early 1960s, Loeb and Sourirajan at UCLA made the critical development of asymmetric cellulose acetate (CA) polymeric membranes, which could be scalably manufactured, and featured a high flux and good rejection desalination process.[4] This crucial milestone has led to the expansion of membrane based separation processes from microfiltration to nanofiltration, reverse osmosis, and transformed membrane based separation from the laboratory to industrial applications.[2,5] Today, along with the emergence of nanotechnology and ever-increasing global challenges, such as clean water, renewable energy and a sustainable environment, great efforts from both industry and academia have brought membrane based separation to a brand new scenario. One recent advance is to use 2D materials, especially graphene and its derivates, for membrane based separation processes. In contrast to polymer membranes, graphene based membranes are still in their early stages of development. As a star material, graphene, since its isolation from bulk graphite in 2004, has been widely explored by a wide range of studies because of its distinct, intriguing electric, thermal, and mechanical properties. However, the study on graphene based membranes was only initiated in 2008, when mechanical exfoliated graphene was proved to be an impermeable atomic membrane.[6] Since then, much research has been directed towards introducing angstrom/nanometer size pores onto graphene sheets and studying their molecular transport phenomena, and many exciting achievements have been made using this protocol, which will be summarized in this book. The other type of graphene based membrane is laminate made from graphene oxide (GO) sheets, which is a 2D material with a graphene backbone decorated with oxygen containing functional groups. The uniform yet angstrom sized channel (pore) structure and the potential scalability promise huge potential for GO membranes. In this chapter, we will review the state-of-art membrane based separation and filtration processes, and in particular, graphene enabled separation processes will be highlighted to align with the theme of this book.

1.2 Membrane Based Separation Processes

Here, seven important membrane based separation processes will be discussed. For the separation of solid from liquid, depending on the size and molecular weight of the retained component (Figure 1.2), the membrane based separation processes can be classified into microfiltration (MF), ultrafiltration (UF), nanofiltration (NF), reverse osmosis (RO), and membrane distillation (MD). In addition, for the separation of two liquids, we will discuss the pervaporation process. Finally, membrane based gas separation will be reviewed.

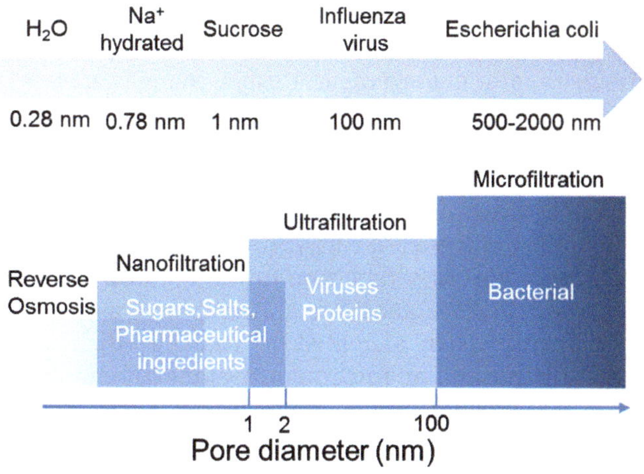

Figure 1.2 Membrane based liquid phase separation technologies.

1.2.1 Microfiltration and Ultrafiltration

Microfiltration is a separation process that removes substances with an effective size of 0.1–10 μm. Currently, the microfiltration process is mainly used for biological testing of bacteria in water, and sterilization, which removes bacteria from water. The membranes are mostly made into disposable cartridges, which are used as the last step to sterilize ultrapure water in the pharmaceutical industry or remove yeast and bacteria for wine and beer.[2] Similar to microfiltration, which relies on the membrane pore size to screen the solute, ultrafiltration membranes provide a pore size in the range of 1–100 nm, which separates solutes of similar size. UF technology is widely used for industrial and municipal water treatment, where bacterial, virus, and colloidal particle screenings are essential steps.[7] The main advantage of the UF process is that, under a low driving pressure (1–5 bar), it offers a reasonably high flux, and, therefore, a high separation efficiency, with a good retention of the solutes. MF and UF are developed technologies, which, like many technologies, still face operational problems, for example, fouling. However, due to their relatively large and open pore structure, the management of the process is rather simpler than that for NF and RO separation processes, which we will discuss later in this chapter.

1.2.2 Nanofiltration

Membranes for NF usually have their pore size in the order of 1 nm, they typically have high rejections to multivalent inorganic salts and small organic molecules and they provide 2 to 5 times higher flux than RO membranes. Depending on the type of solvents, NF separation processes can be aqueous NF or organic solvent nanofiltration (OSN).

1.2.2.1 Aqueous Nanofiltration

Water based NF could find many environmental applications, from desalination to waste water management. Despite many successes made in the field of NF separation, future NF processes still require better contaminant retention and the recovery of valuable components with a high efficiency.[8] Therefore, membranes with higher permeation and/or precise rejection are highly desired. Many nanomaterials could provide nanometer or angstrom sized pores/channels, and such a cavity, if well exploited, should be promising for next-generation NF separation. In the following, taking biological channel-aquaporin, carbon nanotubes and graphene as examples, we discuss recent developments in nanomaterial-enabled efficient NF processes (Figure 1.3a–c). Aquaporin is an integral membrane protein that forms pores in the membrane of biological cells and facilitates water transport between cells. In nature, aquaporin forms ~7 nm long, hourglass shaped channels with a constriction size of 0.28 nm, which can only accommodate one water molecule to pass through in single file (Figure 1.3a). A recent development

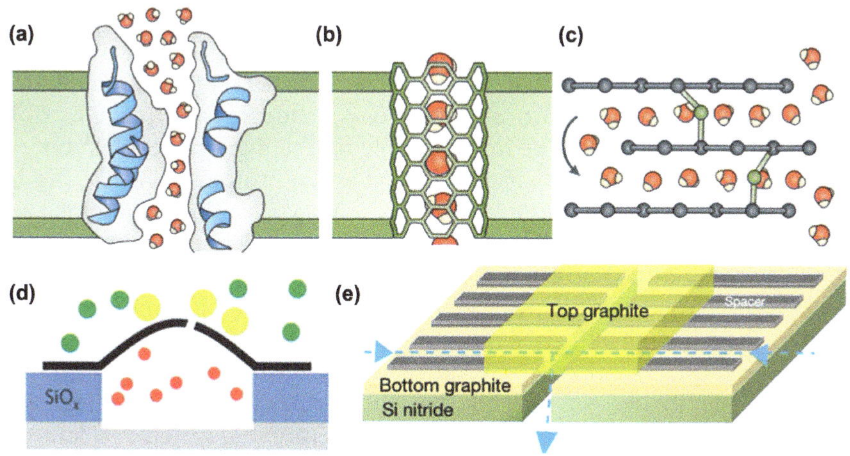

Figure 1.3 Nanomaterial-enabled nanofitration and desalination technologies. (a–c) Water molecules pass through (a) aquaporin, (b) carbon nanotube and (c) GO membrane. (d) a porous mechanical exfoliated graphene membrane. (e) 2D channels made from a Van der Waals heterostructure. Parts a–c reproduced from ref. 33 with permission from Springer Nature, Copyright 2016. Part d reproduced from ref. 17 with permission from Springer Nature, Copyright 2017. Part e reproduced from ref. 18 with permission from Springer Nature, Copyright 2016.

has incorporated aquaporin Z (a bacterial aquaporin from *Escherichia coli*) into copolymer vesicles, and found that the resulting vesicles are up to 800 times higher in water permeability than the copolymer alone while rejecting all the tested salt solutes.[9] The protocol is quite promising as it seems to be an ideal membrane for desalination, however, scalable integration of the aquaporin into a matrix without sacrificing the sieving property, and the mechanical and biological stabilities of aquaporin pose great challenges for its future development.[10] Carbon nanotubes provide a simple, straight and cylindrical channel by seamlessly wrapping their graphitic walls. Synthesis of vertically aligned carbon nanotubes spawned much interest in exploiting carbon nanotubes as nanoscale channels for NF and RO applications (Figure 1.3b).[11–14] Encapsulating the vertical aligned carbon nanotube arrays in silicon nitride made it possible to measure the molecular transport through the carbon nanotube channels. The water permeation was measured to be three to five orders faster than the prediction from continuum flow models.[11,13] However, because the channel size, which is in essence the inner diameter, is >1 nm, the membrane showed Donnan-type rejection, in which the charge interaction between membrane and the ions dominates the ion rejection, therefore, the rejection of mono- and di-valent ions is rather poor.[12] Later on, by inserting 0.8 nm-diameter, 10 nm-long carbon nanotube porins in lipid membrane, the water transport through the carbon nanotube was found to be even higher than that of aquaporins and the carbon nanotube showed a cation/anion selectivity of 184, even in a high ion concentration.[15] Despite the promising permeation and rejection found in carbon nanotubes, scalable and designed integration of carbon nanotubes in the membrane is a great barrier to overcome for carbon nanotube based separation technologies. Intrinsic graphene (made by mechanical exfoliation) sheets were found to be impermeable to tested gas molecules[6] but, as reported by a recent study, permeable to thermal protons.[16] To use the intrinsic graphene as a separation membrane, two types of investigations have been made. One is to introduce angstrom/nanometer sized pores into graphene sheets (Figure 1.3d).[17] The other is to fabricate nanochannels based on 2D heterostructures (Figure 1.3e).[18] The height of the channels, which is the narrowest dimension, is defined by the thickness of the side wall 2D materials. As the pore sizes in the above two scenarios are within size range of the RO membrane, we will discuss this further in RO separation. The other graphene based NF technology is GO membrane, which is a laminar stack of GO sheets and contains 0.9 nm wide nanochannels when soaking in water, enabling precise molecular sieving to ions and molecules. In detail, the membrane allows permeation of species with diameter less than 0.9 nm, but it blocks larger ones.[19] In addition, GO sheets are naturally dispersible in water, therefore, its membrane could be prepared using existing solution based thin film preparation methods, promising a good scalability for large area fabrication. Both the scientific and practical advantages of GO membranes stimulated many investigations on the use of GO membrane for nanofiltration applications.[20–23] Some studies employ GO sheets as an additive in polymer composite membranes to improve the porosity, and enhance the permeance. A more promising method is to use GO based laminar membranes. Based on the foundation of the pioneering discovery of the 0.9 nm cutoff sieving property, current efforts are focused on the fine control of the interlayer distance,[24] which is essentially the pores of the GO membranes, to differentiate ions with diameter less than 0.9 nm, and eventually be able to use the membrane for desalination.

1.2.2.2 Organic Solvent Nanofiltration

OSN has emerged recently as many research and industrial advances have proved it to be an energy and cost efficient solution for many chemical manufacturing industries, for instance, the pharmaceutical, oil refinery, food production industries *etc.*[25–28] Existing organic solvent related separation processes are mostly distillation and evaporation, which account for 80% of operation costs because of the intensive thermal energy and time consumption. In particular, these processes struggle to separate components that are thermally unstable, inducing, for example, thermally induced decomposition or phase changes. OSN could be used for three purposes: (1) concentration: to separate at least one solute from one solvent; (2) solvent exchange: to replace at least one solvent with other solvent(s) with low boiling points, higher affinities to the solute *etc.*; (3) purification, that is, to retain the desired solute by removing other solute(s) from a multi-solute solution.[26] Similar to water based nanofiltration, one of the current research trends of OSN membranes is improving the membrane performance in permeance and selectivity. Ultrathin polymer membranes (normally below 100 nm thin) made by interfacial polymerization with molecular level design of the monomers have achieved a solvent permeance that is 1 to 3 orders of magnitude higher than the state-of-art OSN membranes.[28,29] Another challenge of polymer based OSN membranes is their limited solvent resistance, which undermines the robustness of the separation system. Recent progress on GO based OSN membranes has addressed this issue due to its compatibility with a wide variety of organic solvents.[30,31] Along with the development of a robust, high performance membrane, more efforts on membrane module and process design will speed up the adoption of OSN processes in industrial applications.

1.2.3 Desalination

1.2.3.1 Reverse Osmosis

Reverse osmosis is a process that desalts water using a membrane that favours water permeation but inhibits salt ions under a trans-membrane pressure larger than the osmotic pressure of the salted water. RO separation technology has made significant progress since the development of the Loeb-Sourirajan CA membrane. CA membrane enables technically and economically viable RO by providing high water permeance and a salt rejection >99%.[2] The good performance made CA membrane based RO desalination an industrial standard between the 1960s and mid-1970s. However, CA is susceptible to hydrolysis and requires high pressure for desalination,[5] therefore, interfacial polymerized aromatic polyamide (PA) thin film composite membranes were developed and became the new industry standard from the late 1970s. Today, most of the RO desalination plants use PA membranes.[2,32] In the following four decades, RO technology witnessed many developments of membrane modules to improve the energy efficiency of the desalination process. During that time, many technical efforts have been devoted to improving the membrane's resistance to chlorine, which comes from the water disinfection pretreatment, as PA is prone to decomposing under chlorine attack, which subsequently decreases the lifespan of the membrane and the process. Therefore, many technical efforts, for instance, careful choice of the monomer and surface modification, have been made to improve the membrane chlorine resistance.[32] A major technical challenge for RO membranes, including NF membranes, is membrane fouling,[33,34] that is, the deposition of substances on the membrane surface or inside pores. Membrane fouling originates from the deposition of precipitated salt due to concentration polarization, absorption of organic matter, and biological fouling from deposition and growth of microorganisms. The dense nature, surface charge, hydrophobicity, and surface roughness of PA and most NF membranes make them prone to fouling. To address this issue, altering the surface physicochemical properties by grafting or surface coating,[33] and the process design of cross flow filtration and regular membrane cleansing, have become major current protocols for anti-fouling. It is worth noting that recent reports have shown that PA membrane coated with a thin layer of GO membrane shows a promising low fouling tendency, which is attributed to the hydrophilicity and rather smooth surface of GO membrane.[35–37] Development of fouling control of the RO process is a wide topic and covers many effects from various aspects, readers should refer to recent comprehensive reviews.[38–40]

Although the next-generation of RO membrane has not yet been developed, several nanomaterials have shown promise, for example, zeolites, vertically aligned carbon nanotubes, graphene *etc.* Here, as many have been discussed in the nanofiltration process, graphene is picked to update its research regarding possible RO membranes. For mechanically exfoliated graphene, only recently, 2D channels based on 2 layer graphene/1 layer MoS_2 Van der Waals heterostructures for desalination have been reported (Figure 1.3e). Interestingly, despite a uniform pore size of ~0.66 nm, the 2D channels were found to be permeable to NaCl, the hydrated ionic diameter of which is 0.72 nm. This counterintuitive phenomenon is explained by distortion of the hydrated sodium ions, which could then make it possible to accommodate and transport ions in the smaller channels.[41] Similar results were also found in physically confined GO laminate, in which the ion transport has been studied with a variety of accessible (net) interlayer space from 0.9 to 0.3 nm. It was found that only when the space decreased to ~0.45 nm, could the significant rejection of NaCl be observed, which is explained by the fact that at larger interlayer space, the hydrated ion shrunk by entropy favoured dehydration entered into the channels, and, therefore, showed limited rejection.[24] These two reports implied that, in the case of size extrusion based desalination, a pore size below 0.5 nm would be a possible good route for the desalination process. Interestingly, a recent study supports this implication; desalination has been demonstrated using a porous 5 um diameter graphene membrane fabricated by chemical vapor deposition, in which the pores are generated by short-time plasma treatment (0.5–2 seconds), and it was found that graphene membrane with a pore size in the range of 0.5–1 nm provides a H_2O and KCl selectivity of ~100.[42] The consistent results for three different types of graphene further confirm the importance of pore size for graphene based desalination. However, in practice, such uniform, fine pores would be a great challenge to fabricate and upscale as many attempts have been made but only with limited success. Fabrication innovation in this matter would be an important direction for graphene based desalination membranes.

1.2.3.2 Membrane Distillation

Two alternative desalination technologies, forward osmosis (FO) and membrane distillation (MD), which could utilize low-grade waste energy, have become a popular theme of study under the current global sustainable energy challenges. FO separation involves, instead of applying energy-consuming physical pressure, extracting water from the salted water into a draw solution by its chemical osmotic pressure and further separating the draw solute and water from the diluted draw solution. Readers should refer to several comprehensive reviews to understand more aspects of this technology.[43,44] In the following, we discuss the MD separation process. MD separates pure water from salted water by driving a vapor pressure gradient through a porous hydrophobic membrane. Parallel flows of hot feed solution and cold/room temperature permeate flow are separated by the membrane, inside which water vapor instead of liquid water is transported from the feed to permeate side. As water is vaporized and transports to the permeate side through Knudsen diffusion, salt ions that could not be vaporized are impermeable, therefore the MD process in principle could achieve 100% rejection to salted water. The other advantage of the MD process is it allows the use of an alternative energy source, for example, solar energy. However, the process comes with some drawbacks, for example, low permeation flow, relatively high heat lost by conduction and fouling.[45–47] Recently, nanocarbon materials, due to their good electrical and photothermal conversion efficiency, have been used to address some of these challenges.[48–50] One type of these materials is the carbon nanotube; carbon nanotubes have been employed as a highly efficient joule heater to directly heat up the permeate salt solution under alternating current at the membrane interface, and the resulting MD module was able to reach 100% single-pass recovery, which even exceeds the theoretical prediction for standard MD processes.[48] Another method is using carbon black nanodots as an efficient photothermal heater, which enables direct utilization of solar illumination power to drive the distillation process, and this achievement largely extended MD distillation as a promising solution for household or community scale desalination.[49] Graphene is also used as an anti-fouling interface for MD processes, and this property is explained by the weak physical absorption of contaminants on the graphene surface.[50]

1.2.4 Pervaporation

Different to the above mentioned liquid based separation technologies, which involve mostly solutions containing solid solutes and solvents, the pervaporation process separates solvent mixtures: the mixture is heated on the feed side with contact to the membrane, and the permeate side is vacuumized by either a pump or condensation induced vacuum. Pervaporation could be found in niche applications, mostly for organic solvent dehydration, water purification from trace organic solvents, and organic/organic solvent separation as an alternative to distillation. Solvent dehydration is a major application for the pervaporation process.[2] During the production of alcohols, an azeotrope of alcohol and water forms, which makes distillation not applicable and makes pervaporation a good choice.

A recent finding on the water permeable but organic solvent impermeable property of GO membranes has raised a lot interest in using GO membrane as an ideal pervaporation material.[23,51] In future, with the rise of biofuels, for example, ethanol and butanol, pervaporation separation may witness a development in the market.

1.2.5 Gas Separation

Gas separation has become a very important industrial application of membrane technology in the past 40 years, as it was intensively involved in fuel production, CO_2 emission and industrial gas purifications.[52,53] Bonded with its application background, important gas pairs are mostly energy- and CO_2- related, for example, N_2/O_2, CO_2/N_2, H_2/CH_4, CO_2/CH_4, C_2H_4/C_2H_6, C_3H_6/C_3H_8.[53] Like RO membranes, almost all gas separation membranes are dense membranes; the gas transport through the membrane follows a solution-diffusion model, therefore, the selectivity is determined by both the sorption selectivity and diffusion selectivity. Leveraging between the sorption and diffusion selectivity to enable an efficient separation process is a great challenge.[53] Currently, as well as polymer based membranes, nanomaterial based membranes and their composite membranes have emerged. For example, zeolite, metal organic framework, and carbon molecular sieving. These molecular sieves, with potential tunability of their chemistry, offer a variety of opportunities to obtain high sorption selectivity. On the other hand, their rigid pores (comparing to polymer based membranes) with engineering possibility could greatly alter the diffusion selectivity of gas pairs with similar size. These synergetic advantages enabled a new generation of gas separation membranes.[53] Except the above nanomaterials, graphene has also been intensively studied for gas separation. The research focus is mostly on the porous mechanically exfoliated graphene and the GO membranes. By introducing angstrom sized pores on the mechanically exfoliated graphene, the resulting graphene membrane could achieve very high selectivity over the gas molecule pairs.[17] Despite the fact that, as mentioned in the same scenario for graphene based liquid separation processes, there is a long way ahead for the practical application, researchers, in the short-term, are using a simple yet neat platform to study the molecular transport, which, in turn, will promote the understanding and development of gas separation technologies. GO laminar based gas separation membranes have been intensively studied since the publication of two individual reports on ultrathin GO membranes for efficient H_2 separation.[54,55] Different to most dense gas separation membranes, the preparation of ultrathin GO membrane, especially, with only several layer thinness, requires extra care to control the defects or assembly voids to ensure efficient gas separation.

In this chapter, state-of-art membrane separation technologies have been reviewed. It is interesting to conclude that, despite its centuries-long history, only in the past five to six decades have significant advances been made in this field, that is to say, membrane separation, with the emergence of nanomaterials, for example graphene, and advances of nanofluidics, is still and will be in its rapid development stages for a long time and many exciting novel membrane materials, processes, and applications can be envisioned.

References

1. D. S. Sholl and R. P. Lively, *Nature*, 2016, **532**, 435.
2. R. W. Baker, *Membrane Technology and Applications*, Wiley Online Library, 2012.
3. J. A. Nollet, *J. Membr. Sci.*, 1995, **100**, 1.
4. S. Loeb and S. Sourirajan, *Saline Water Conversion—II*, American Chemical Society, 1963.
5. E. M. V. Hoek and V. V. Tarabara, *Encyclopedia of Membrane Science and Technology*, Wiley Online Library, 2013.
6. J. S. Bunch, S. S. Verbridge, J. S. Alden, A. M. van der Zande, J. M. Parpia, H. G. Craighead and P. L. McEuen, *Nano Lett.*, 2008, **8**, 2458.
7. E. Nagy, in *Handbook of Water and Energy Management in Food Processing*, Woodhead Publishing, 2008.
8. A. W. Mohammad, Y. H. Teow, W. L. Ang, Y. T. Chung, D. L. Oatley-Radcliffe and N. Hilal, *Desalination*, 2015, **356**, 226.
9. M. Kumar, M. Grzelakowski, J. Zilles, M. Clark and W. Meier, *Proc. Natl. Acad. Sci. U. S. A.*, 2007, **104**, 20719.
10. D. M. Stevens, J. Y. Shu, M. Reichert and A. Roy, *Ind. Eng. Chem. Res.*, 2017, **56**, 10526.
11. J. K. Holt, H. G. Park, Y. Wang, M. Stadermann, A. B. Artyukhin, C. P. Grigoropoulos, A. Noy and O. Bakajin, *Science*, 2006, **312**, 1034.
12. F. Fornasiero, H. G. Park, J. K. Holt, M. Stadermann, C. P. Grigoropoulos, A. Noy and O. Bakajin, *Proc. Natl. Acad. Sci. U. S. A.*, 2008, **105**, 17250.
13. M. Majumder, N. Chopra, R. Andrews and B. J. Hinds, *Nature*, 2005, **438**, 44.
14. M. Whitby and N. Quirke, *Nat. Nanotechnol.*, 2007, **2**, 87.
15. R. H. Tunuguntla, R. Y. Henley, Y.-C. Yao, T. A. Pham, M. Wanunu and A. Noy, *Science*, 2017, **357**, 792.
16. S. Hu, M. Lozada-Hidalgo, F. C. Wang, A. Mishchenko, F. Schedin, R. R. Nair, E. W. Hill, D. W. Boukhvalov, M. I. Katsnelson, R. A. W. Dryfe, I. V. Grigorieva, H. A. Wu and A. K. Geim, *Nature*, 2014, **516**, 227.
17. L. Wang, M. S. H. Boutilier, P. R. Kidambi, D. Jang, N. G. Hadjiconstantinou and R. Karnik, *Nat. Nanotechnol.*, 2017, **12**, 509.
18. B. Radha, A. Esfandiar, F. C. Wang, A. P. Rooney, K. Gopinadhan, A. Keerthi, A. Mishchenko, A. Janardanan, P. Blake, L. Fumagalli, M. Lozada-Hidalgo, S. Garaj, S. J. Haigh, I. V. Grigorieva, H. A. Wu and A. K. Geim, *Nature*, 2016, **538**, 222.
19. R. K. Joshi, P. Carbone, F. C. Wang, V. G. Kravets, Y. Su, I. V. Grigorieva, H. A. Wu, A. K. Geim and R. R. Nair, *Science*, 2014, **343**, 752.
20. J. Y. Chong, B. Wang and K. Li, *Curr. Opin. Chem. Eng.*, 2016, **12**, 98.

21. A. Gugliuzza, A. Politano and E. Drioli, *Curr. Opin. Chem. Eng.*, 2017, **16**, 78.
22. K. Huang, G. Liu, Y. Lou, Z. Dong, J. Shen and W. Jin, *Angew. Chem., Int. Ed.*, 2014, **53**, 6929.
23. G. Liu, W. Jin and N. Xu, *Chem. Soc. Rev.*, 2015, **44**, 5016.
24. J. Abraham, K. S. Vasu, C. D. Williams, K. Gopinadhan, Y. Su, C. T. Cherian, J. Dix, E. Prestat, S. J. Haigh, I. V. Grigorieva, P. Carbone, A. K. Geim and R. R. Nair, *Nat. Nanotechnol.*, 2017, **12**, 546.
25. P. Marchetti, L. Peeva and A. Livingston, *Annu. Rev. Chem. Biomol. Eng.*, 2017, **8**, 473.
26. P. Marchetti, M. F. Jimenez Solomon, G. Szekely and A. G. Livingston, *Chem. Rev.*, 2014, **114**, 10735.
27. M. G. Buonomenna and J. Bae, *Sep. Purif. Rev.*, 2014, **44**, 157.
28. M. F. Jimenez-Solomon, Q. Song, K. E. Jelfs, M. Munoz-Ibanez and A. G. Livingston, *Nat. Mater.*, 2016, **15**, 760.
29. S. Karan, Z. Jiang and A. G. Livingston, *Science*, 2015, **348**, 1347.
30. Q. Yang, Y. Su, C. Chi, C. T. Cherian, K. Huang, V. G. Kravets, F. C. Wang, J. C. Zhang, A. Pratt, A. N. Grigorenko, F. Guinea, A. K. Geim and R. R. Nair, *Nat. Mater.*, 2017, **16**, 1198.
31. L. Huang, J. Chen, T. Gao, M. Zhang, Y. Li, L. Dai, L. Qu and G. Shi, *Adv. Mater.*, 2016, **28**, 8669.
32. S. S. Shenvi, A. M. Isloor and A. F. Ismail, *Desalination*, 2015, **368**, 10.
33. J. R. Werber, C. O. Osuji and M. Elimelech, *Nat. Rev. Mater.*, 2016, **1**, 16018.
34. M. A. Shannon, P. W. Bohn, M. Elimelech, J. G. Georgiadis, B. J. Marinas and A. M. Mayes, *Nature*, 2008, **452**, 301.
35. F. Perreault, M. E. Tousley and M. Elimelech, *Environ. Sci. Technol. Lett.*, 2014, **1**, 71.
36. S. Bano, A. Mahmood, S.-J. Kim and K.-H. Lee, *J. Mater. Chem. A*, 2015, **3**, 2065.
37. X. Huang, K. L. Marsh, B. T. McVerry, E. M. V. Hoek and R. B. Kaner, *ACS Appl. Mater. Interfaces*, 2016, **8**, 14334.
38. J. H. Jhaveri and Z. V. P. Murthy, *Desalination*, 2016, **379**, 137.
39. D. Rana and T. Matsuura, *Chem. Rev.*, 2010, **110**, 2448.
40. G.-d. Kang and Y.-m. Cao, *Water Res.*, 2012, **46**, 584.
41. A. Esfandiar, B. Radha, F. C. Wang, Q. Yang, S. Hu, S. Garaj, R. R. Nair, A. K. Geim and K. Gopinadhan, *Science*, 2017, **358**, 511.
42. S. P. Surwade, S. N. Smirnov, I. V. Vlassiouk, R. R. Unocic, G. M. Veith, S. Dai and S. M. Mahurin, *Nat. Nanotechnol.*, 2015, **10**, 459.
43. T. Y. Cath, A. E. Childress and M. Elimelech, *J. Membr. Sci.*, 2006, **281**, 70.
44. S. Zhao, L. Zou, C. Y. Tang and D. Mulcahy, *J. Membr. Sci.*, 2012, **396**, 1.
45. A. Alkhudhiri, N. Darwish and N. Hilal, *Desalination*, 2012, **287**, 2.
46. K. W. Lawson and D. R. Lloyd, *J. Membr. Sci.*, 1997, **124**, 1.
47. A. Deshmukh, C. Boo, V. Karanikola, S. Lin, A. P. Straub, T. Tong, D. M. Warsinger and M. Elimelech, *Energy Environ. Sci.*, 2018, DOI: 10.1039/c8ee00291f.

48. A. V. Dudchenko, C. Chen, A. Cardenas, J. Rolf and D. Jassby, *Nat. Nanotechnol.*, 2017, **12**, 557.
49. P. D. Dongare, A. Alabastri, S. Pedersen, K. R. Zodrow, N. J. Hogan, O. Neumann, J. Wu, T. Wang, A. Deshmukh, M. Elimelech, Q. Li, P. Nordlander and N. J. Halas, *Proc. Natl. Acad. Sci. U. S. A.*, 2017, **114**, 6936.
50. D. H. Seo, S. Pineda, Y. C. Woo, M. Xie, A. T. Murdock, E. Y. M. Ang, Y. Jiao, M. J. Park, S. I. Lim, M. Lawn, F. F. Borghi, Z. J. Han, S. Gray, G. Millar, A. Du, H. K. Shon, T. Y. Ng and K. Ostrikov, *Nat. Commun.*, 2018, **9**, 683.
51. R. R. Nair, H. A. Wu, P. N. Jayaram, I. V. Grigorieva and A. K. Geim, *Science*, 2012, **335**, 442.
52. P. Bernardo, E. Drioli and G. Golemme, *Ind. Eng. Chem. Res.*, 2009, **48**, 4638.
53. W. J. Koros and C. Zhang, *Nat. Mater.*, 2017, **16**, 289.
54. H. W. Kim, H. W. Yoon, S. M. Yoon, B. M. Yoo, B. K. Ahn, Y. H. Cho, H. J. Shin, H. Yang, U. Paik, S. Kwon, J. Y. Choi and H. B. Park, *Science*, 2013, **342**, 91.
55. H. Li, Z. Song, X. Zhang, Y. Huang, S. Li, Y. Mao, H. J. Ploehn, Y. Bao and M. Yu, *Science*, 2013, **342**, 95.

CHAPTER 2

Graphene-based Membranes

GONGPING LIU AND WANQIN JIN*

Department of Chemical Engineering, State Key Laboratory of Materials-oriented Chemical Engineering, Nanjing Tech University, 5 Xinmofan Road, Nanjing 210009, China
*Email: wqjin@njtech.edu.cn

2.1 Introduction

Membranes are a key component in systems such as batteries, packages, controlled delivery, solar cells and large-scale energy-efficient separation and purification processes.[1] In the last category, an engineered membrane controls the sorting of components by selective passage between feed and permeate streams. Such a membrane separation technology plays an important role in reducing the costs and environmental impacts of the production of commodities with an order of magnitude less energy requirement compared with traditional approaches. Highly permeable and selective membranes are required to enable wider application of membrane technology.[2]

Polymeric membranes have dominated the practical application of membrane technology, while they generally suffer a trade-off between permeability and selectivity.[3] For inorganic membranes, although exhibiting high separation performance, challenges remain in cost-efficiency or scalable fabrication. On the other hand, carbon-based materials show great potential in mass transport owing to their well-defined nanostructures and low-friction surface. In the past decade, carbon nanotubes (CNTs) have been considered as ideal promising building blocks for high-performance membranes because of the unique one-dimensional (1D) nanochannel for

Nanoscience & Nanotechnology Series No. 47
Graphene-based Membranes for Mass Transport Applications
Edited by Pengzhan Sun and Hongwei Zhu

Published by the Royal Society of Chemistry, www.rsc.org

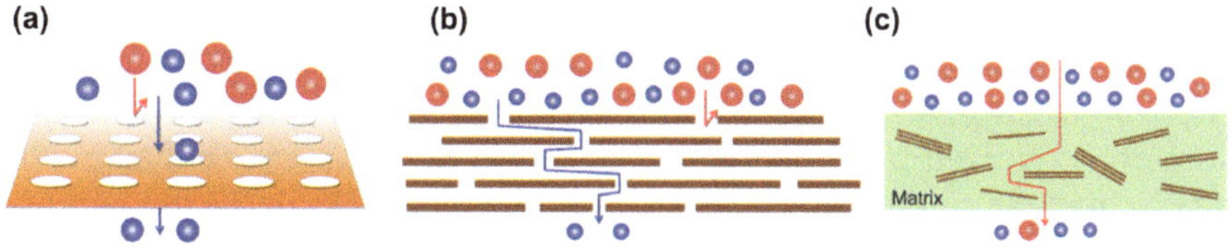

Figure 2.1 Main types of graphene-based membranes: (a) porous graphene membranes; (b) graphene laminate membranes; (c) graphene hybrid membranes.

ultra-fast molecular transport.[4] However, the high-cost of CNTs and complex process of obtaining high-density and large-area vertically-aligned CNTs limit CNT-based membranes to basically theoretical studies only.

Since the 2010 Nobel Prize for "ground-breaking experiments regarding the two-dimensional (2D) material graphene",[5] graphene-based materials have been used for developing a new family of membranes with extraordinary separation properties.[6] The atomic thickness and nearly frictionless surface of 2D graphene is able to provide the resulting membrane with minimum transport resistance and maximum permeate flux. The outstanding mechanical strength and chemical stability, along with the cost-effective production of graphene materials, offer additional advantages in terms of practical application. According to the microstructures, graphene-based membranes can be classified into three types:[7] (1) porous graphene membranes, (2) graphene laminate membranes and (3) graphene-based hybrid membranes, as shown in Figure 2.1. This chapter will discuss the progress in these graphene-based membranes for separation, including the principles, design, fabrication, application and transport mechanism.

2.2 Porous Graphene Membranes

2.2.1 Perfect Single-layer Graphene

Ideally, a graphene sheet can be used as the ultimate membrane for separation. The single carbon atom thick graphene could maximize the membrane flux since membrane permeance is generally inversely proportional to the membrane thickness. However, perfect single-layer graphene is impermeable to molecules, because the π-orbitals of graphene form a dense and delocalized cloud that blocks the gap within its aromatic rings. Theoretical calculation[8] indicated that there is no space in electron density around the aromatic rings for passing molecules. The geometric pore size of graphene is 0.064 nm, which is smaller than the smallest molecule helium with a kinetic diameter of 0.28 nm. This impermeability of graphene sheets was experimentally demonstrated by Bunch and co-workers.[9] They tested the permeation of several gases through a graphene sheet mechanically exfoliated from graphite by the Scotch tape technique. The monolayer graphene membrane was found to be impermeable to gases as small as helium. Also, the excellent mechanical strength of graphene enabled the single-layer graphene membrane to withstand pressure differences as high as 6 bar. These impermeable and robust properties of graphene sheets motivated numerous theoretical and experimental studies on perforated graphene sheets, namely porous graphene membranes, for selectively transporting water, gases and ions.

2.2.2 Molecular Simulation Studies

Initially, molecular simulations were used to study the molecular transport through porous graphene membranes. Cohen-Tanugi and Grossman[10] used classical molecular dynamics to investigate water desalination across nanoporous single-layer graphene with the function of pore size, functional groups bonded to the pore edge, and applied pressure (Figure 2.3). They found that salt rejection is strongly dependent on the pore size. Moreover, graphene pores functionalized by hydroxyl groups can almost double the water flux owing to their hydrophilic character, as well as the ability of hydroxyl groups to substitute for water molecules in the hydration shell of the ions. The optimized nanoporous single-layer graphene membrane exhibits a water permeability of 2750 $L\,m^{-2}\,h^{-1}\,bar^{-1}$ that is 2–3 orders of magnitude higher than conventional reverse osmosis membranes with a similar salt rejection of 99% (Figure 2.2). With this great promise, they further explored the desalination performance of multilayer nanoporous graphene membranes[11] that can be synthesized more economically than the single-layer material. Compared with the same thickness as a polyamide membrane (~200 nm), 200-layer nanoporous graphene membrane shows a water permeability of ~2 $L\,m^{-2}\,h^{-1}\,bar^{-2}$, which is on the same order of magnitude as that of a typical polyamide membrane. This suggests that the ultrathin layer might be the main advantage of nanoporous graphene membrane.

Sint *et al.*[12] designed functionalized nanopores in graphene monolayers for selective ion passage by molecular dynamics simulations. They built graphene pores with a diameter of ~5 Å and various functionalized atoms. It was found that ions can pass through these pores as long as their hydration shells can be partly removed. The charge effect dominated the ion passage through the graphene nanopore. Specifically, a negatively charged F–N terminated pore allows the passage of cations (Li^+, Na^+, K^+) while blocking the anions. By contrast, a positively charged H terminated pore allows the passage of anions (F^-, Cl^-, Br^-) but it rejects cations. These interesting results demonstrated the feasibility of nanoporous graphene membranes for ion separation involved in water purification and energy storage systems.

Porous graphene membrane for gas separation was initially simulated by Jiang and co-workers[13] by using first principles density functional theory calculations. The gas selectivity was determined by the computed diffusion barriers. The designed N-functionalized pore with a diameter of 3.0 Å shows high H_2 permeance and H_2/CH_4 selectivity on the order of 10^8. By further reducing the pore diameter to 2.5 Å by building an all-hydrogen passivated

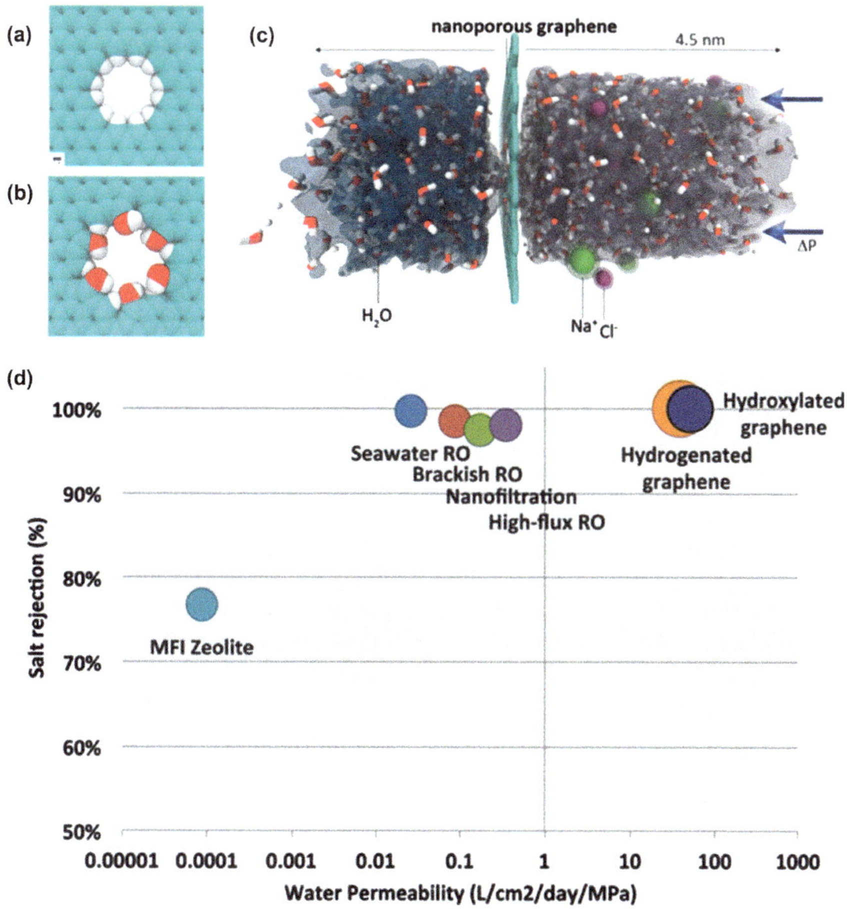

Figure 2.2 Molecular simulation on porous graphene membranes: (a) hydrogenated and (b) hydroxylated graphene pores; (c) side view of the computational system; (d) performance chart for functionalized nanoporous graphene *versus* existing technologies.
Reprinted with permission from D. Cohen-Tanugi and J. C. Grossman, *Nano Letters*, 2012, **12**, 3602–3608, Copyright 2012 American Chemical Society.[10]

pore, extremely high selectivity on the order of 10^{23} for H_2/CH_4 can be achieved. Apparently, these membrane pores are far superior to traditional polymer and silica membranes based on bulk solubility and diffusivity dominating the transport of gas molecules. Nevertheless, precisely controlling the pore size of graphene membranes down to 3–5 Å remains challenging for current perforation techniques. Thus, Tian *et al.*[14] proposed an ion-gated graphene membrane comprising a monolayer of [emim][BF_4] ionic liquid-coated porous graphene monolayer to reduce the pore size from the order of 1 nm to 3–4 Å to achieve selective gas separation. They demonstrated that this composite membrane is able to show excellent CO_2/CH_4

separation performance with CO_2/CH_4 selectivity of ~42 and CO_2 permeance of ~10^5 GPU [1 GPU = 10^{-6} cm^3 (STP)/(cm^2 s cmHg)]. Additionally, the gas separation performance can be tuned by the anion size, pore size, and ionic liquid thickness.

2.2.3 Experimental Studies

In simulation studies, well-defined nanopores with a high number density can be ideally created in monolayer graphene. Although precisely controlling pore size, high pore density, and large-area perforation remains challenging, extensive efforts have been devoted to this purpose.[15] Table 2.1 summarizes the advantages and limitations of different perforation methods for fabricating nanoporous graphene membranes.[16] Bunch and co-workers[17] demonstrated ultraviolet-induced oxidative etching for creating Angstrom scale pores in micrometer-sized graphene membranes. The transport of gases (H_2, CO_2, Ar, N_2, CH_4 and SF_6) with different molecular size through the nanoporous graphene bilayer was measured by a previously reported pressurized blister test and mechanical resonance.[9] The obtained separation performance partially agreed with models based on effusion through a small number of angstrom-sized pores. This work is considered as a notable step toward the realization of macroscopic, size-selective porous graphene membranes.

Another experimental breakthrough[18] realized a nanoporous single-layer graphene membrane for water desalination. As a convenient method, an oxygen plasma etching process was used to fabricate and tailor the nanopores with a precise size of 0.5–1 nm in suspended single-layer graphene (Figure 2.3). The resulting nanopores showed excellent rejection of ions such

Table 2.1 Advantages and limitation of different perforation methods for fabricating porous graphene membranes. Reproduced from ref. 16 with permission from the American Chemical Society, https://pubs.acs.org/doi/abs/10.1021/acs.jpclett.5b00914, Copyright 2015.

Method	Pore size (nm)	Advantage	Limitation
Focused electron beam ablation	~3.5	Tunable and well-defined pore size	Limited to small area
Ultraviolet-induced oxidative etching	<10	Can treat large-area samples	Low pore density wide size distribution
Ion bombardment and chemical oxidative etching	<1	Can treat large-area samples Tunable pore size	Moderate pore density
Oxygen plasma etching	~1	Can treat large-area samples Tunable pore size	Moderate pore density
Focused ion beam irradiation	5–100	Tunable and well-defined pore size	Limited to small area pores too large for salt rejection

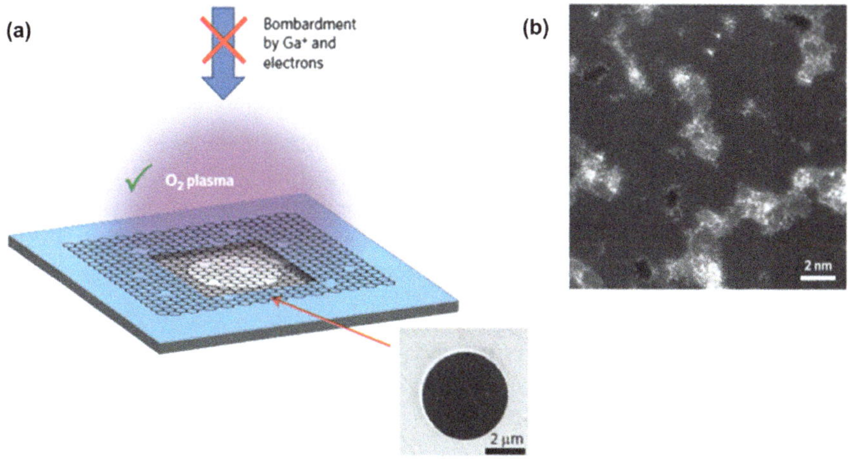

Figure 2.3 Experimental results for porous graphene membranes: (a) schematic and SEM image of single-layer graphene suspended on a 5-µm-diameter hole. Several approaches have been used for nanoporous graphene fabrication: bombardment by ions, by electrons and *via* oxygen plasma treatment; (b) aberration-corrected scanning transmission electron microscopy (STEM) images of graphene after 1.5 s exposure to oxygen plasma.
Reprinted by permission from Springer: Springer Nature, *Nature Nanotechnology*, Water desalination using nanoporous single-layer graphene, S. P. Surwade, S. N. Smirnov, I. V. Vlassiouk, R. R. Unocic, G. M. Veith, S. Dai and S. M. Mahurin, Copyright 2015.[18]

as K^+, Na^+, Li^+ and Cl^-. The estimated water flux through the single nanopore can reach 3 molecule/ps, which is 3 orders of magnitude higher than that through aquaporin channels and 1 order of magnitude greater than the values estimated by molecular dynamics simulations. While in a diffusion process with an osmotic pressure gradient, the porous graphene membrane showed a smaller water flux of 200 molecules/ms that is ~5 times lower than fluxes through an aquaporin channel and 4 times less than the estimation from molecular dynamics simulations.

To address the challenge of producing a high density of pores and large-area perforation, Karnik and co-workers[19] used ion bombardment to introduce isolated and reactive defects on single-layer graphene, and subsequently applied oxidative etching to enlarge these defects to nanopores with diameters of 0.40 ± 0.24 nm and densities of $>10^{12}\,cm^{-2}$. Oxidation time was used to tune the pore size. With short oxidation time, the created small pores were cation-selective because of the electrostatic repulsion function, while a longer oxidation time generated larger pores, allowing salt passage but rejecting organic molecules based on a steric size exclusion effect. Alternatively, a CVD growth followed by focused ion beam approach[20] was developed to fabricate bilayer nanoporous graphene membranes with pore sizes ranging from <10 nm to 1 µm and a large number of pores

($\sim 10^3$ to 10^6 per membrane). The resulting atomically thin graphene membrane was applied for practical permeation measurements, showing orders-of-magnitude higher permeance for water and gases compared with the state-of-the-art membranes. In addition, the realization of nanoporous graphene membranes is limited by leakage through defects in the ultrathin graphene. These defects are inevitably formed during the CVD growth and transfer process of graphene layer. A multiscale leakage-sealing process developed by Karnik's group[21] might provide a feasible approach of constructing defect-tolerant nanoporous graphene membranes for water purification.

2.3 Graphene Laminate Membranes

2.3.1 Fabrication Methods

Despite the extraordinary separation performance that has been predicted and somehow achieved in nanoporous graphene membranes, fabrication of large-area single/few-layered graphene membranes with uniform nanopores remains a challenge.[16,22] Alternatively, graphene laminate membranes assembled from graphene nanosheets provide a more practical approach for application of graphene-based membranes.[19] Interlayer channels between graphene nanosheets can offer fast and selective transport of small molecules. Graphene oxide (GO),[23] a derivative of graphene, is often used to construct the graphene laminate membranes because of the following features: (1) GO is similar to graphene in that it also exhibits single-atom thickness with lateral dimensions as high as tens of micrometers; (2) GO contains plenty of oxygen-containing groups, such as hydroxyl, carboxyl and epoxy groups on the edges and basal planes, making it easily dispersed in water solution; (3) GO can be massively and cost-effectively produced *via* chemical oxidization and ultrasonic exfoliation of graphite. To improve the stability of laminate membranes in water, GO can be partially reduced, which is termed rGO. Thus, graphene laminate membranes are mostly related to GO or rGO membranes.

Filtration is the most common and straightforward route for the fabrication of GO membranes. Ruoff and co-workers[24] fabricated freestanding GO papers by filtration assembly of GO sheets. A tightly packed interlocking network of sheets was formed that subtly undulate along the paper surface. They proposed that the laminar structure is mainly controlled by (1) slowly flowing water in confined galleries and (2) electrostatic and van der Waals attractive forces between the GO nanosheets. Typically, the thickness of the GO membranes can be tuned by the filtrated amount of GO nanosheets that is often varied by the concentration and volume of the GO suspension. A porous substrate is often used to fabricate a thin-film composite GO membrane, whose pore size, porosity and surface roughness affect the formation of the uniform GO membrane. Surface modification of the substrate by introducing some functional groups can improve the interfacial adhesion between GO membrane and substrate.[25]

Hung and co-workers[26] studied the driving force of the filtration on the nanostructures of GO laminate membranes. As shown in Figure 2.4a, two typical filtration processes, pressure-assisted and vacuum-assisted filtration, were compared with the evaporation-assisted method. It was found that the self-assembly technique with evaporation generated a heterogeneous GO layer with loop structures. However, more ordered laminar structures were obtained by using pressure-assisted and vacuum-assisted filtration, since the transmembrane pressure can be easily controlled by the pressure-assisted process. These morphologies were further confirmed by the *d*-spacings of the GO membranes prepared by pressurized filtration, vacuum filtration and evaporation method varying from 8.3 Å to 11.5 Å, respectively. Moreover, both GO lateral size and stacked layer thickness were found to be important for preparing high-quality GO membranes.[27] Different lateral sizes of GO nanosheets can be obtained by controlling the sonication and centrifugation process, and the thickness of the GO membrane can be reduced by dilution of GO dispersions.

Coating, including drop-casting,[28] dip-coating,[25] spin-coating,[29] and spray-coating,[30] has been used to fabricate graphene laminate membranes. Kim *et al.*[29] used two coating methods to prepare few-layered GO membranes with a thickness of 3–10 nm: (1) drop-casting followed by spin-coating and (2) spin-casting (Figure 2.4b). They found that the stacking of GO nanosheets is governed by GO edge-to-edge intrinsic repulsive interactions and face-to-face attractive capillary forces created by the spin-coating. The drop-casting process formed a relatively heterogeneous GO deposition caused by the electrostatic repulsion between the GO edges. The spin-casting induced a highly interlocked laminate because the capillary interactions between the faces of GO sheets overcome the electrostatic forces between the GO edges. Jin and co-workers[31] studied a spray-coating followed by solvent evaporation process for tuning the microstructure of GO laminate membranes. It was indicated that disordered-to-ordered and porous-to-compact GO membrane structures can be finely and conveniently manipulated *via* controlling the spraying times and evaporation rate during GO assembly. Compared with other coating methods, the application of the spray-coating approach is not limited by the configuration of substrates, not only flat sheet, but also tubular or even hollow fiber forms can be applied.

Layer-by-layer (LBL) assembly was also used to assemble GO nanosheets into laminate membranes. The thickness of GO membrane can be precisely controlled by varying the number of layer-by-layer deposition cycles. Moreover, this approach is ideal for uniformly introducing molecules or nanomaterials during each layer deposition to functionalize or intercalate GO membranes. When integrated with spin-coating, the LBL method shows higher efficiency and structural homogeneity for preparing GO membranes. Hu and Mi[32] applied GO nanosheets as water separation membranes *via* this method, as shown in Figure 2.4c. They introduced 1,3,5-benzenetricarbonyl trichloride (TMC) during the LBL process to crosslink the GO laminates. The interlayer galleries of the resulting GO laminates were intercalated by

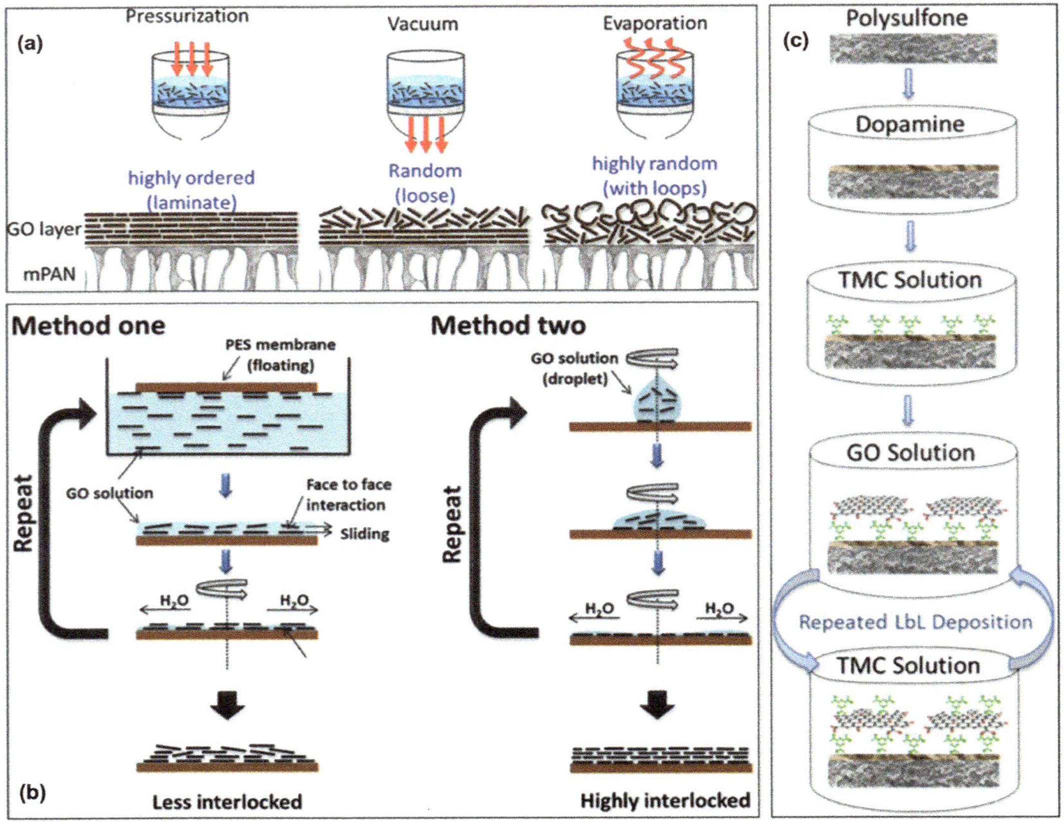

Figure 2.4 Schematic of graphene laminate membranes prepared by different methods: (a) pressure-assisted self-assembly, vacuum-assisted self-assembly, and evaporation-assisted self-assembly; (b) spinning coating; and (c) layer-by-layer. Reproduced from ref. 7 with permission from the Royal Society of Chemistry.

the TMC molecules (0.7 nm), leading to an estimated interlayer spacing of ~1 nm. The crosslinking process was considered to have two functions: (1) regulating the charges, functionality and spacing of the GO nanosheets and (2) stabilizing the stacked GO nanosheets to overcome their inherent dispensability in the water environment. Hung *et al.*[33] used different diamines (ethylenediamine, butylenediamine, or *p*-phenylenediamine) to crosslink GO membranes for ethanol/water separation. The covalent bonding between the GO nanosheets and diamine could effectively suppress the swelling of the GO membranes in water, thereby enhancing the membrane separation performance and long-term stability.

Scalable fabrication is practically important for the development of GO membranes. Akbari *et al.*[34] proposed a scalable and industrially adaptable method to fabricate large-area (13×14 cm^2) graphene laminate membranes within 5 s by shear-induced alignment of liquid crystals of GO. The highly ordered GO nanosheets in the plane of the laminate membrane led to organized channels and enhanced water permeability. Jin and co-workers[35] developed rGO membrane on the inner surface of ceramic tubes *via* a pressurized filtration method. The tubular GO/ceramic composite membranes exhibited excellent mechanical strength, as well as cross-flow and high-pressure filtration performance when nanoparticles were *in situ* grown on the graphene nanosheet. By optimizing the filtration process, the tubular GO membrane was successfully scaled up from 5 to 40 cm length. Further, it is expected that this kind of membrane could be easily applied for depositing GO membrane on multi-channel ceramic substrates, highly improving the packing density and reducing the cost of GO membranes.

2.3.2 Tuning Nanostructures

There are at least two kinds of transport channels in graphene laminate membranes: (1) plane-to-plane interlayer galleries formed by face-to-face GO nanosheet interaction and (2) in-plane structural defects and/or slit-like pores formed by edge-to-edge GO nanosheet interaction. The transport of molecules through a GO laminate membrane occurs *via* in-plane slit-like pores and then plane-to-plane intergalleries.[36] A great number of studies demonstrated that the interlayer galleries between GO nanosheets play a vital role in the fast and selective transport of water, ions and gases. Thus, how to precisely tune the interlayer spacing is key to achieving high-performance GO membranes. Mi summarized several methods to obtain interlayer galleries with tunable size corresponding to various applications for separation.

The interlayer galleries in GO laminate membranes are controlled by the attached oxygen-containing functional groups, face-to-face interactions and corrugation of GO nanosheets.[37] Geim and co-workers[38] reported that the empty interlayer spacing of a typical GO membrane is ~0.5 nm. The pristine interlayer spacing was enlarged when the GO membrane was immersed in water or other solvents, because of the hydration[38] or solvation[39] of oxygen-containing groups on the GO nanosheets. For instance, water-swelled GO membranes exhibited an empty interlayer spacing of more than ~0.9 nm. Usually, the interlayer spacing is measured by using XRD, however measuring *d*-spacing up to several nanometers in an aqueous solution presents a practical technical challenge for XRD measurement. Mi and co-workers[40] developed an integrated quartz crystal microbalance with dissipation and ellipsometry approach to quantify the *d*-spacing of GO membranes in liquids that went well beyond the typical measurement limit of ~2 nm.

On the other hand, the interlayer galleries can be narrowed by removing the oxygen-containing groups on GO *via* thermal or chemical routes. Once the interlayer spacing was significantly decreased, the GO membrane even exhibited a barrier property,[41] namely it was highly impermeable to all gases and liquids. Proper reduction of GO could supress the redispersion of GO laminates so as to improve the structural stability of the GO membrane in water. However, it is challenging to carry out an *in situ* reduction treatment for GO membrane while maintaining the membrane integrity. Alternatively, GO nanosheets can be firstly reduced into rGO nanosheets dispersed in solution, then the rGO dispersion is used for fabricating rGO membranes using the conventional methods for preparing GO membranes.[42] For example, GO was reduced with L-ascorbic acid (L-AA) in water solution where the C–O–C and –OH groups on GO nanosheets were reduced to C=C groups by L-AA, and L-AA was oxidized into D-AA.[43] The resulting rGO nanosheets were still dispersed well in water, likely due to the fact that L-AA could form a shell around the edges of the reduced GO sheets to prevent aggregation.

Intercalation is a widely used approach to enlarge the interlayer spacing of graphene laminate membranes. Different kinds of nanomaterials and molecules have been intercalated into GO membranes, showing improved separation performance. Owing to the good compatibility with GO, carbon-based materials such as carbon dots,[44] single-walled carbon nanotubes (SWCNTs)[45] and multi-walled carbon nanotubes[46] are preferred to be used as intercalating materials. By properly expanding the interlayer space between adjacent GO nanosheets, the intercalated GO laminate membranes exhibited 1–2-fold water flux improvements while maintaining the same rejection rate.[44–46] Organic molecules such as diamine[33] were also intercalated into the GO laminates to create crosslinked GO frameworks with improved structural stability. In the meantime, the size of the molecules can be tuned to control the interlayer spacing of the intercalated GO membranes.

In some cases, although water permeance was enhanced by intercalation, uncontrollable and uneven intercalated species inevitably disturbed the pre-stacked GO nanosheets, thereby remaining insufficient to afford high efficiency (selectivity). Therefore, it's important to find a way of maximally expanding the pristine channels of GO laminates to boost the water permeance without sacrificing the membrane selectivity. Jin and co-workers[35] proposed a general and scalable approach involving nanoparticles

(NPs)@rGO as building blocks for laminate membranes to realize this purpose. In this approach, NPs (such as Fe_3O_4, UiO-66, TiO_2) were uniformly *in situ* grown on the plane of GO nanosheets, thereby effectively enlarging the interlayer galleries to maximize the permeance while withstanding the disordered laminar structure to preserve the membrane selectivity. Simultaneously, GO is partially reduced to rGO, offering additional structural stability in water. More importantly, the intercalated rigid NPs along with the formed NPs–rGO bonding also provided the NPs@rGO membrane with sufficient stability for surviving the high-pressure and cross-flow filtration process. It was further demonstrated that intercalation of nanoporous nanoparticles like UiO-66 MOFs could provide additional transport channels,[47] leading to higher water permeance than that of intercalating nonporous Fe_3O_4 nanoparticles with the same particle size.

The intercalated nanomaterials can be removed to create more transport channels. Huang *et al.*[48] mixed positively charged $Cu(OH)_2$ nanostrands with negatively charged GO nanosheets to form tightly intercalated GO membranes *via* electrostatic attraction. A highly porous network of nanochannels with a narrow size distribution of 3–5 nm was formed by dissolving the $Cu(OH)_2$ nanostrands in an acid solution, leading to a 10-fold enhancement in water permeance. Unfortunately, these nanochannels, without rigid supports within the interlayer galleries, were found to easily collapse when applying a few bars of filtration pressure, and the water permeance dropped remarkably. A similar phenomenon was also found in another 2D material laminate membrane.[49]

Regardless of the substantial progress in GO-based membranes showing high water flux, the rejection rates for ions, especially monovalent ions, of most GO-based membranes were still lower than the commercial polymeric membranes for nanofiltration[46] or reverse osmosis. Controlling the membrane pore size within the subnanometer range is required for precisely sieving monovalent ions from water during desalination processes but this is challenging for graphene laminate membranes. To address this challenge, Jin and co-workers[50] rationally designed external forces to precisely manipulate the GO laminates with an empty interlayer height of ~0.4 nm. As shown in Figure 2.5, the external forces (EFs) applied outside the GO laminate (so-called "outer" EFs) include compressive force, centrifugal force and shear force, which were applied by using a facile vacuum-spin coating process. The EFs applied inside the GO laminate (so-called "inner" EFs) refer to GO – polymer molecular interactions, which were provided by intercalating PEI molecules during each LBL cycle. The "outer" and "inner" EFs performed collaboratively to overcome the intrinsic repulsive electrostatic forces between GO nanosheets, thereby eliminating non-selective stacking defects and stimulating a highly ordered assembly of GO nanosheets. The GO laminate membrane induced by only "outer" EFs had an initial *d*-spacing of ~0.76 nm. After further introducing "inner" EFs by intercalating an appropriate number of polyetherimide (PEI) molecules, the *d*-spacing was gradually enlarged to 0.76 nm with well-maintained

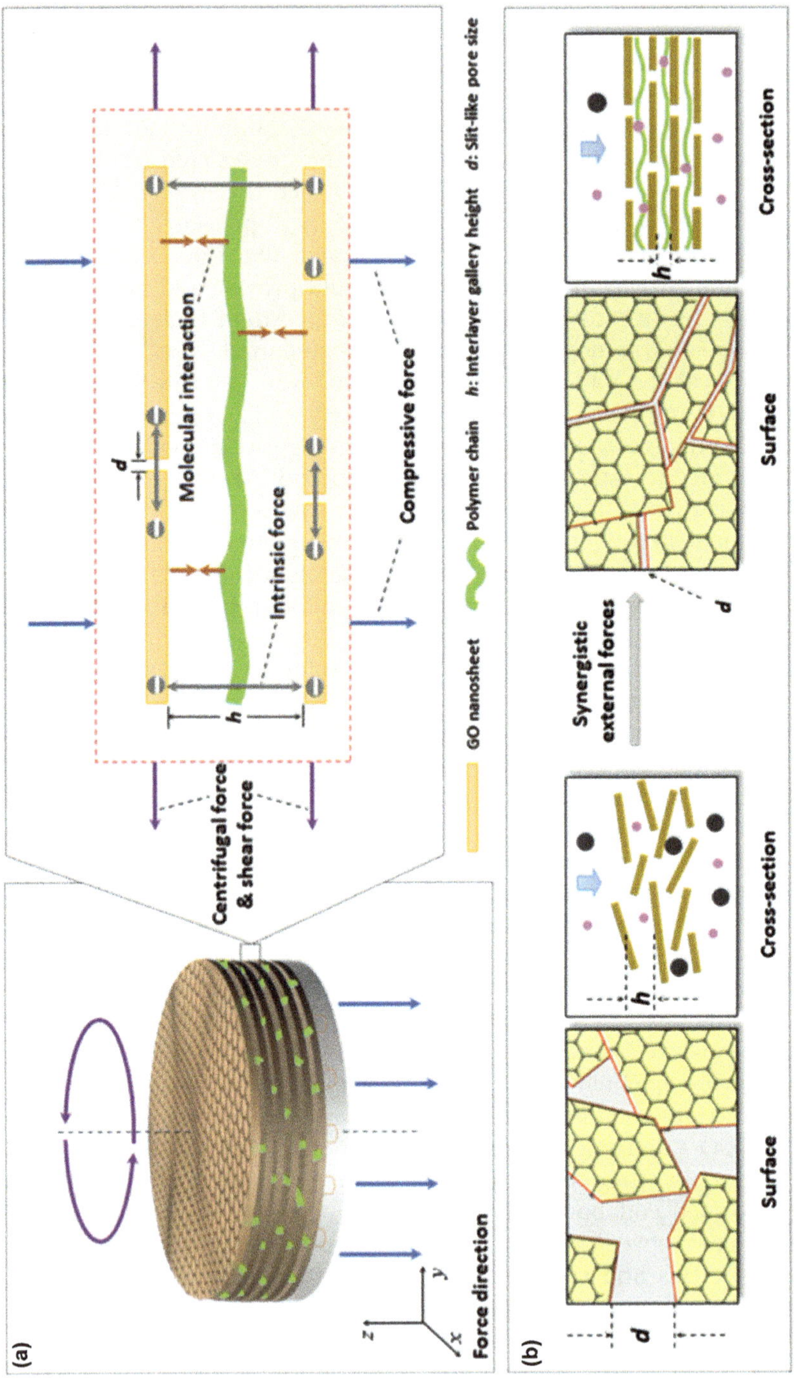
(a)
z
y
x
Force direction
Centrifugal force
& shear force
Molecular interaction
d
h
Intrinsic force
Compressive force
GO nanosheet
Polymer chain
h: Interlayer gallery height
d: Slit-like pore size
(b)
d
h
Surface
Cross-section
Synergistic
external forces
d
h
Surface
Cross-section

crystallinity. Considering the graphene thickness of 0.34 nm, the height of empty space between GO nanosheets is calculated to be 0.42 nm, which is suitable for sieving ions and gases.

As discussed above, the application of GO membranes in ion sieving and desalination technologies is limited by a permeation cut-off of ~9 Å[38] caused by the water swelling, which is larger than the diameter of hydrated ions of common salts. To achieve smaller interlayer spacing for the GO membranes immersed in water, Nair and co-workers[51] proposed physical confinement to control the interlayer spacing of GO laminates from ~9.8 Å to 6.4 Å that matches the required size for sieving hydrated ions. Specifically, stacked GO laminates were embedded in epoxy, which mechanically restricts the swelling of the laminate on exposure to RH or liquid water. The GO–epoxy stacks were glued into a slot made in either a metal or plastic plate, then two sides of these stacked membranes were trimmed off to open the nanochannels. Thus, ions and water molecules permeate along the laminate direction rather than vertically to the laminate in most permeation studies on graphene laminate membranes.

In addition to molecular diffusion through interlayer channels and pores or defects of graphene layers, the sorption on the surface of graphene laminate membranes also contributes to the selective transport of species. For instance, solvent dehydration performed by GO membranes is governed by a sorption–diffusion model.[52] According to the quartz crystal microbalance measurement, water molecules are preferentially adsorbed onto the membrane surface and are then transported through the GO interlayer galleries across the membrane. A recent study[46] indicated that the water sorption process might be the control step for the solvent dehydration process. This means that the interlayer galleries might be not fully utilized for the separation if the surface adsorption is insufficient. To enhance the surface sorption property, Jin and co-workers[53] designed a bio-inspired membrane by depositing an ultrathin surface water-capturing polymeric layer (<10 nm) on top of GO laminates. Chitosan (CS) was chosen as the

Figure 2.5 Design and construction of 2D channels. (a) External force driven assembly approach for fabricating 2D channels. This involves three-dimensional external forces in the *x*, *y*, and *z* directions. Enlarged schematic shows force analysis for one 2D channel unit consisting of GO nanosheets and polymer chain. Three main types of forces are included: intrinsic force, "outer" external forces (compressive force, centrifugal force and shear force), which are applied outside the 2D channel unit and "inner" external forces (GO – polymer molecular interactions), which are applied inside the 2D channel unit. (b) Hypothetical evolution of surface and cross-section of GO-assembled 2D channels from an intrinsic force induced disordered structure (left) to highly ordered laminar structures (right) driven by introduced synergistic external forces.
Reprinted with permission from J. Shen, G. Liu, K. Huang, Z. Chu, W. Jin and N. Xu, *ACS Nano*, 2016, **10**, 3398–3409, Copyright 2016 American Chemical Society.[50]

hydrophilic layer to promote the preferential capture of water molecules from feed mixtures, thereby increasing the driving force across the GO laminates. As a result, the interlayer galleries for water transport could be used fully to achieve highly improved water flux through the integrated membrane.

2.3.3 Separation Applications

Since Nair *et al.*[30] reported the exciting finding of unimpeded water permeation through GO laminates, increasing attention has been paid to developing graphene-based membranes for water purification. Table 2.2 summarizes the water separation performance of GO laminate membranes. GO membranes readily prepared by filtration, LBL or coating methods have been intensively studied regarding the filtration processes of ultrafiltration (UF), nanofiltration (NF) and forward osmosis (FO). To obtain sufficient water flux and mechanical strength, thin GO laminate membranes with sub-micrometer or few-micrometer thickness were deposited on top of porous substrates such as polycarbonate (PC), polyvinylidene fluoride (PVDF), polysulfone (PSF) and poly(acrylonitrile) (PAN), as well as ceramics. The membrane configurations varied from flat sheet to tubular and hollow fiber.

The reported GO membranes are efficient for rejecting organic dyes such as direct yellow (DY), Evans blue (EB) and methylene blue (MB) with the rejection rate of 85%–100%, while they are much less efficient for rejections of monovalent or divalent salts (rejection rate <50%). This is mainly due to the GO membranes in water showing a swelled interlayer spacing of $\geq\sim 1$ nm, which is within the molecular size of organic dyes but too large for retaining salts. Creating more transport channels, either by intercalating nanomaterials or increasing porosity, has been proven to be effective for improving the water permeance. The nanostrand channeled GO membranes showed 1 order of magnitude higher water permeance than that of the pristine GO membranes.[48] With the intercalation of 10 nm diameter MWCNTs into graphene laminate membrane, the water permeability reaches 52.7 $L\,m^{-2}\,h^{-1}\,bar^{-1}$, which is ~5 times that of pristine rGO membrane and 5–10 times higher than most commercial NF membranes.[54] By *in situ* growing nanoparticles (NPs) on GO nanosheets, the resulting NPs@rGO membranes filtrated on the inner surface of a ceramic tube exhibited water permeance as high as ~300 $L\,m^{-2}\,h^{-1}\,bar^{-1}$ with excellent rejections for various organic matter and heavy metal ions,[16] as shown in Figure 2.6. Moreover, this new kind of industrially preferred GO-based tubular membrane showed an extraordinary structural stability under a transmembrane pressure up to 15 bar and cross-flow process of water purification, which is considered as a notable step toward realizing scalable GO-based membranes.

Negatively charged groups on GO nanosheets could afford electrostatic interaction with the charged molecules or ions in nanofiltration and salt rejection.[36] According to the exclusion theory, the potential at the interphase

Table 2.2 Graphene-based membranes for pressure filtration.

Graphene-based membranes	Preparation method	Membrane structure[a]	Feed system	Membrane process	Water flux ($L/m^2/h/bar$)	Rejection	Ref.
GO/PC	Vacuum filtration	$l = 1.9$ μm	15 μM EB	UF 0.1 MPa	7.1	85%	48
Nanostrand channeled GO/PC	Vacuum filtration	$d = 3$–5 nm $l = 2$ μm	15 μM EB	UF 0.1 MPa	69.5	83.5%	48
Base-refluxing reduced GO/	Vacuum filtration	$d = 1$–2 nm $l = 22$ nm	0.02 mM MB	NF 0.1 MPa	21.8	99.2%	36
PVDF	Vacuum filtration	$l = 53$ nm	20 mM Na_2SO_4 20 mM NaCl	NF 0.5 MPa	3.3	60% 40%	36
TMC crosslinked GO/PSF	Layer-by-layer	$d = 1$ nm	7.5 mg L^{-1} MB 20 mM Na_2SO_4 20 mM NaCl	NF 0.34 MPa	8–27.6	46–66% 26–46% 6–19%	32
PAH crosslinked GO/PAN	Layer-by-layer	$d = 1$ nm	1 M sucrose	FO 0.34 MPa	2.1–5.8	~99%	32
rGO-80MWCNTs(10)/PC	Pressure-assisted filtration	$d = 0.96$ nm $l = 1230$ nm	10 ppm MB 0.03 mM AO7 0.02 mM RhB	NF 0.1 MPa	52.7	99.8% 99.4% 100%	54
rGO-MWCNTs(2:1)/PVDF	Pressure-assisted filtration	$l = 40$ nm	0.01M Na_2SO_4 0.02 mM DY 50 ppm MO	NF 0.5 MPa	11.33	81.0% 99.8% 96.1%	46
rGO-CNTs/AAO	Vacuum filtration	$l = 600$ nm	50 ppm MO 0.005M Na_2SO_4	NF 0.3 MPa	32 28	97.3% 84%	42
UiO-66-rGO-0.7/PC	Pressure-assisted filtration	$l = 300$ nm	10 ppm RhB	NF 0.2 MPa	14.4	98.2%	47
Fe_3O_4@rGO/Ceramic support	Pressure-assisted filtration	$l = 10$ μm	50 ppm RhB	NF 0.2 MPa	296	98.14%	35

[a] *d*: pore size; *l*: membrane thickness; PAI: poly(amide-imide); PEI: polyethyleneimine; TMC: 1,3,5-benzenetricarbonyl trichloride; PAH: poly(allylaminehydrochloride); EB: Evans blue; MB: methylene blue; RhB: rhodamine B, DY: direct yellow.

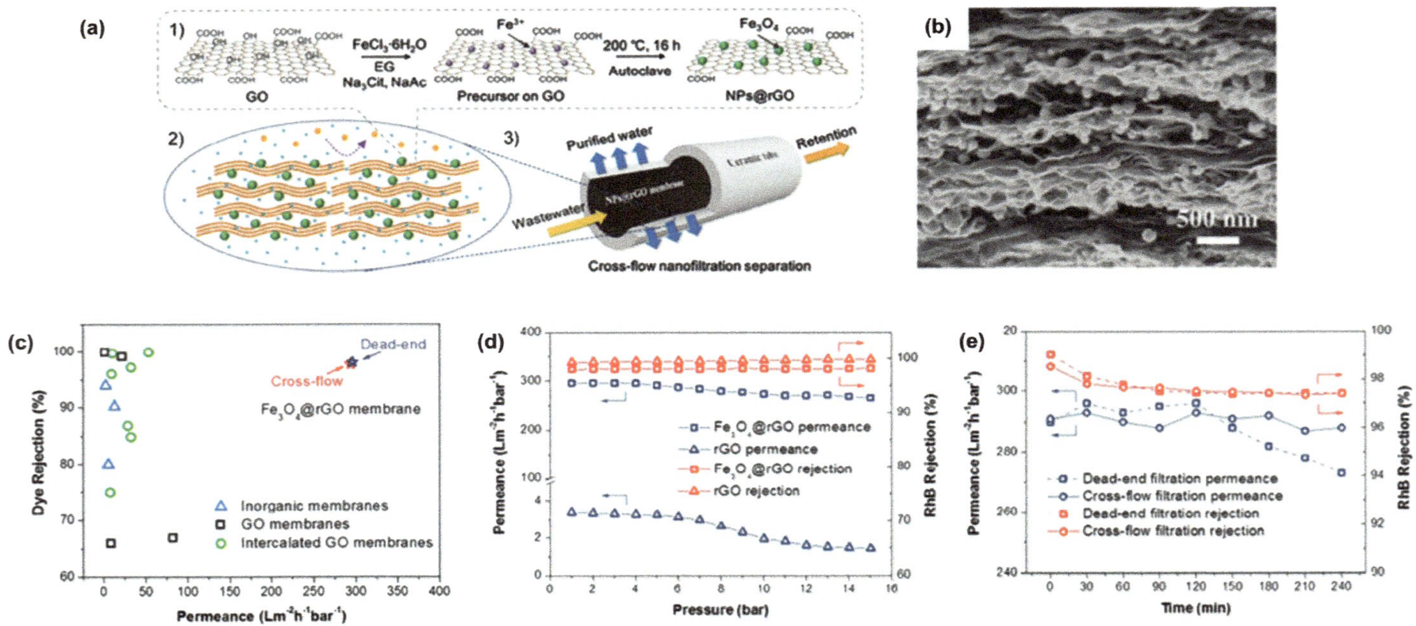

Figure 2.6 Graphene-based membranes for pressure filtration: (a) illustration of (1) *In situ* synthesis of NPs@rGO nanosheets, (2) the assembled laminar structure of NPs@rGO membrane, and (3) cross-flow nanofiltration of the membrane deposited on the inner surface of a ceramic tube; (b) SEM images of cross-sectional morphology of Fe_3O_4@rGO membrane; (c) Performance comparison with reported relevant membranes (inorganic, GO, intercalated GO membranes for dyes rejections; Water permeance and RhB rejection of rGO and Fe3O4@rGO membranes under (d) feed pressures from 1 to 15 bar and (e) dead-end/cross-flow operation (for Fe3O4@rGO membrane only). (Feed: 50 ppm RhB; Pressure: 2 bar; Temperature: 25 °C). Reproduced from ref. 35 with permission from John Wiley and Sons, © 2017 American Institute of Chemical Engineers.

of solution and membrane tends to exclude co-ions, and counter ions are rejected to maintain the solution electro-neutrality on each side of the membrane. Zhu and co-workers[28,55] carried out systematic studies on selective ion transport through GO laminar membranes in which diverse interactions were formed between different ions and different regions (*e.g.*, *sp*^{2} clusters and oxygen-containing functional groups) of GO. The diverse interplays between ions and GO result in different interaction strengths, and the balance between ion–GO interactions and the hydration–dehydration effect of ions lead to excellent selectivity of the GO membranes toward various species in solutions.

Another application of water transport through graphene laminate membranes is solvent dehydration *via* the pervaporation process. In a typical pervaporation process, feed solution is passed over the membrane surface and some of the components preferentially pass through the membrane and are concentrated as vapors in the vacuumed permeate side. This process follows the solution-diffusion model. Owing to the intrinsic hydrophilic property of GO nanosheets, GO membranes have been widely used for selectively removing water from aqueous solutions containing concentrated organic solvents such as ethanol (EtOH), isopropyl alcohol (IPA) and *n*-butanol (1-BtOH). The performance is listed in Table 2.3. Hollow fiber is an industrially preferred membrane configuration because of its advantages in high-packing density and cost-effective fabrication. GO nanosheets were stacked on a ceramic hollow fiber *via* the facile filtration method,[52] showing excellent selective water permeation for pervaporation separation of dimethyl carbonate/water mixtures. As shown in Figure 2.7, by coating an ultrathin hydrophilic CS layer, the prepared CS@GO/ceramic composite membrane exhibits highly improved water permeation with an excellent water flux of over 10 000 $g\,m^{-2}\,h^{-1}$, which exceeds the performance upper bound of state-of-the-art membranes for butanol dehydration.[53]

Graphene is an ideal starting material for developing gas separation membranes because of its atomic thickness, high mechanical strength and impermeability to gases. Porous graphene membranes and graphene laminate membranes were applied for gas separation such as purifying hydrogen or carbon dioxide from gas mixtures. Although extremely high selectivity in the order of 10^{23} for H_2/CH_4 was predicted through the porous graphene monolayer with a pore size of 2.5 Å,[13] experimental perforated bilayer graphene sheets, however, demonstrated much lower selectivity.[17] On the other hand, graphene laminate membranes, prepared by a spin coating or filtration approach, exhibited high gas separation performance. Either H_2[13,17,27,29,38] or CO_2[29,56] would preferentially permeate through the few-layered GO membranes, which was determined by the microstructures of GO laminates. The gas separation performance of these membrane is listed in Table 2.4.

The polar groups such as –COOH and –OH on GO have favorable interactions with C–O bonds in CO_2, showing preferential adsorption for CO_2 over CH_4, N_2 and H_2.[27] Despite this fact, it was interesting to find that

Table 2.3 Graphene-based membranes for pervaporation.

Graphene-based membranes	Preparation method	Membrane structure[a]	Feed condition	Total flux ($g\ m^{-2}h^{-1}$)	Separation factor	Ref.
EDA-crosslinked GO/CA	Pressure-assisted self-assembly	ds = 0.93 nm l = 412 nm	80 °C, 90 wt% EtOH-water	2297	4491	33
GO/mPAN	Pressure-assisted self-assembly	ds = 0.85 nm l = 231 nm	70 °C, 70 wt% IPA-water	4137	1164	64
			70 °C, 90 wt% 1-BtOH-water	4340	1791	26
GO/ceramic hollow fiber	Vacuum suction	l = 1 μm	25 °C, 97.4 wt% DMC/water	1702	743	52
Freestanding GO laminate	Pressurized ultrafiltration	l = 2 μm	24 °C, 75 wt% EtOH-water	1300	211	33
CS@GO/ceramic hollow fiber	Vacuum suction	l = 1 μm	70 °C, 90 wt% 1-BtOH-water	10124	1523	53
GO/ceramic disk	Dip-coating	l = 0.7–1.5 μm	40 °C, 90 wt% EtOH-water	462	13	25
GO/ceramic tube	Pressurized filtration	ds = 0.84 nm l = 1.3 μm	70 °C, 90 wt% EtOH-water	1200	250	65
GO/TFNC	Spin-coating or vacuum filtration	N/A	70 °C, 80 wt% EtOH-water	2200	308	66
GO/AAO	Vacuum filtration	l = 1–3 μm	50 °C, 90 wt% 1-BtOH-water	3100	230	67
GO-sodium alginate composite	Film casting	l = 1.6 μm	76 °C, 95 wt% EtOH-water	1699	1566	58

[a] *ds*: *d*-spacing; *l*: membrane thickness; CA: cellulose acetate; mPAN: modified PAN: polyacrylonitrile; DMC: dimethyl carbonate.

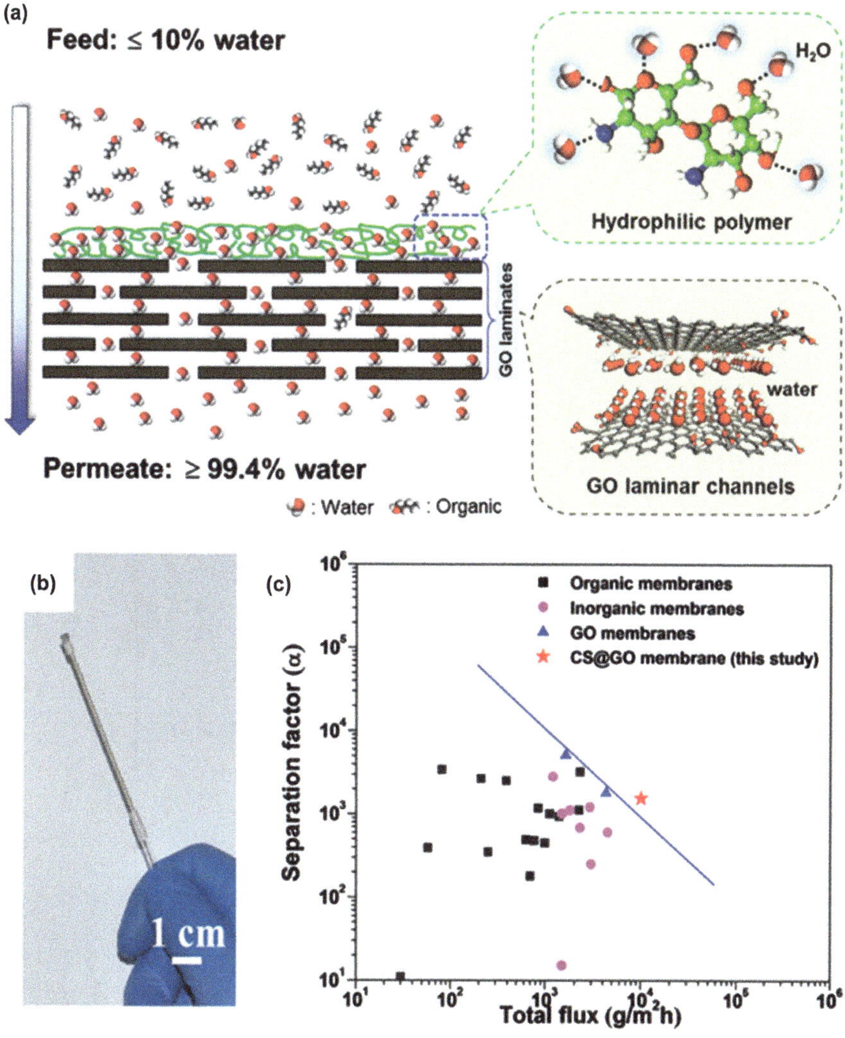

Figure 2.7 Graphene-based membranes for pervaporation: high-efficiency water-transport channels using the synergistic effect of a hydrophilic polymer and GO laminates: (a) schematic of water–organic separation process using the synergistic effect of hydrophilic polymer and GO laminates, (b) digital photograph of prepared CS@GO membrane, (c) comparison of GO membranes and CS@GO membrane with state-of-the-art membranes for water/butanol dehydration.
Reproduced from ref. 53 with permission from John Wiley and Sons, © 2015 WILEY-VCH Verlag GmbH & Co. KGaA, Weinheim.

few-layered GO membranes prepared by different stacking methods exhibited either H_2-selective (H_2/CO_2 selectivity = 30) or CO_2-selective permeation properties (CO_2/N_2 selectivity = 20), respectively.[29] Besides the interlayer galleries, the in-plane structural defects and/or slit-like pores

Table 2.4 Graphene-based membranes for gas separation.

Graphene-based membranes	Preparation method	Membrane structure[a]	Feed condition[b]	Permeate rate/permeance/permeability[c]	Selectivity	Ref.
Few-layered graphene/PTMSP	Chemical vapor deposition	$l = 5$ graphene layers	O_2/N_2	29 Barrer	6	29
Few-layered GO/PES	Spinning coating	$l = 3–7$ nm	CO_2/N_2	100 GPU	20	29
	Thermal reduced	Not available	H_2/CO_2 (140 °C)	1000 GPU	40	29
GO/AAO	Vacuum filtration	$l = 9$ nm	H_2/CO_2	10^{-7} mol m^{-2} s^{-1} Pa^{-1}	3400	27
			H_2/N_2		900	
GO-PEI/Al_2O_3 substrate	Layer-by-layer	$l = 1$ μm	H_2/CO_2	1200 Barrer	29	50
			H_2/C_3H_8		260	
GO/Al_2O_3 substrate	Spray-evaporation	$l = 1$ μm	H_2/CO_2	2.7×10^{-8} mol m^{-2} s^{-1} Pa^{-1}	20.9	31
GO/Al_2O_3 hollow fiber	Vacuum filtration	$l = 300$ nm	H_2/CO_2	1.3×10^{-7} mol m^{-2} s^{-1} Pa^{-1}	15	68
GO-borate/PES	Vacuum filtration	$l = 8$ nm	CO_2/CH_4	650 GPU	75	57
			CO_2/N_2	700 GPU	60	
GO/AAO	Spinning coating	$l = 20$ nm	H_2/CO_2	3.4×10^{-7} mol m^{-2} s^{-1} Pa^{-1}	240	69
GO-PEBA composite	Film casting	$l = 5$ μm	CO_2/N_2	100 Barrer	91	56
ZIF-8@GO composite	Layer-by-layer	$l = 100$ nm for GO layer	H_2/CO_2	1.3×10^{-7} mol m^{-2} s^{-1} Pa^{-1}	15	62
			H_2/N_2		91	
			H_2/CH_4		139	
			H_2/C_3H_8 (250 °C)		3817	
ZIF-8/GO composite	Secondary Growth	$l = 100$ nm ZIF-8/GO layer	H_2/N_2	9.5×10^{-8} mol m^{-2} s^{-1} Pa^{-1}	8.9	63

[a] *d*: pore size; *l*: membrane thickness.
[b] operated at room temperature unless otherwise indicated.
[c] 1 GPU $= 10^{-6}$ cm^3 (STP)/(cm^2 s cmHg); 1 Barrer $= 10^{-10}$ cm^3 (STP) cm/(cm^2 s cmHg). PTMSP: poly(1-methylsilyl-1-propyne); PES: polyether sulfone; AAO: anodic aluminum oxide.

could also contribute or even dominate the gas transport. Li *et al.*[27] used a facile filtration process to fabricate GO membranes as thin as 1.8 nm, showing surprisingly high selectivity for H_2/CO_2 (3400) and H_2/N_2 (900) mixtures. The GO membranes, after reduction, showed 3 orders of magnitude lower water flux, while gas permeance remained unchanged. This suggested that the main transport pathway and permeation of gas molecules was attributed to selective structural defects rather than interlayer spacing.

To achieve substantial gas sieving properties, both the plane-to-plane interlayer galleries and in-plane slit-like pore require subnanometer size. By introducing external forces to manipulate the ordered stacking of GO nanosheets, the resulting GO membranes showed remarkable molecular-sieving properties with 2–3 orders of magnitude higher H_2 permeability and a 3-fold enhancement in H_2/CO_2 selectivity compared with commercial membranes[50] (Figure 2.9a–c). Ultrathin GO membranes using borate as a crosslinker of GO nanosheets, meanwhile, facilitated transport of CO_2 for CO_2 capture.[57] The resulting membranes exhibited excellent CO_2/CH_4 separation performance with high CO_2 permeance up to 650 GPU and CO_2/CH_4 selectivity of 75 (Figure 2.8a–c).

2.4 Graphene Hybrid Membranes

2.4.1 Graphene Oxide/Polymer Hybrid Membranes

Like the other nanomaterials, 2D graphene materials can be combined with polymers to form hybrid membranes.[56,58,59] This approach integrates the high performance of the 2D graphene filler with the good processibility of the polymer matrix, providing an effective way to improve the separation performance of the polymeric membrane. With abundant oxygen-containing groups such as hydroxyl, carboxyl, and epoxide groups on the basal plane and edge, GO is highly dispersible in solution and has good compatibility with various polymers.

Owing to their extraordinary water transport properties and hydrophilicity, GO nanosheets were used as a high-performing 2D filler for pervaporation dehydration of organic/water mixtures. Cao *et al.*[58] incorporated GO nanosheets into sodium alginate (SA) polymer to fabricate GO/SA hybrid membranes. The structural defects, edge-to-edge slit-like pores and non-oxide regions of GO could provide fast and selective transport channels for water molecules. Their study suggested that a smaller lateral size, more structural defects and fewer oxygen-containing groups in the GO nanosheets might afford more channels for water transport, thereby enhancing the water flux of the hybrid membrane. The hybrid of GO and polymer is also a promising approach to enhance the performance of proton exchange membranes. Chen and coworkers[60] fabricated sulfonic acid functionalized GO/Nafion hybrid membranes for proton transport, showing a 4 times higher proton conductivity over Nafion at 120 °C with 25% humidity.

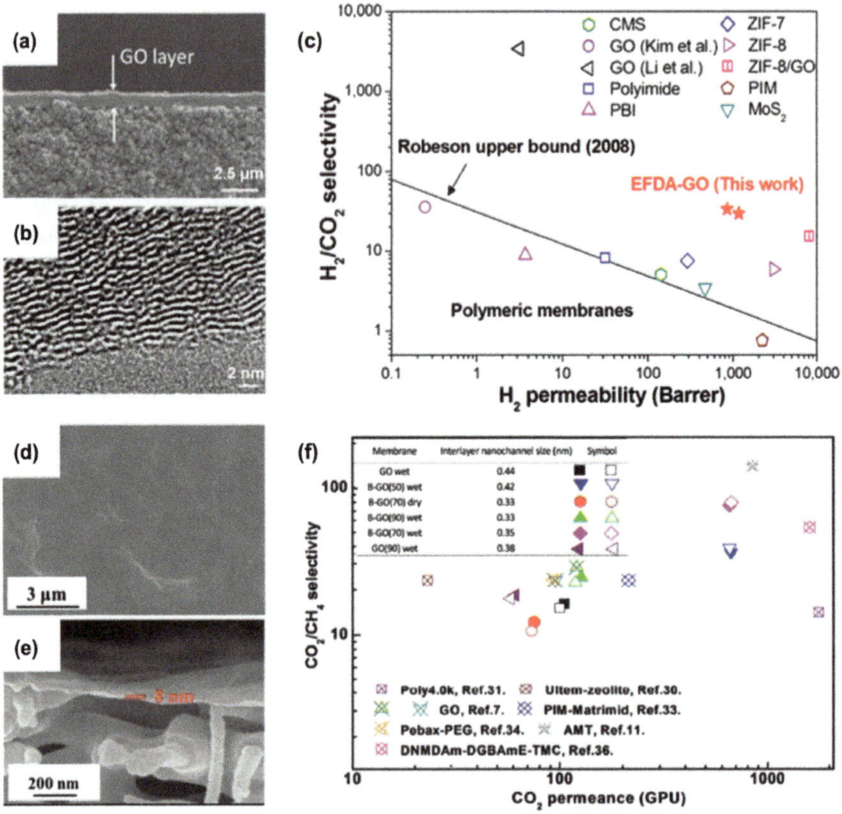

Figure 2.8 Graphene-based membranes for gas separation: (a–c) external force driven assembled (EFDA) GO membrane: (a) SEM image of the membrane cross-section; (b) TEM image of the ordered GO laminar structure; (c) H_2/CO_2 separation performance compared with state-of-the-art gas separation membranes. (d–f) Ultrathin GO membrane: SEM images of (d) the surface and (e) cross-section; (f) CO_2/CH_4 separation performance (solid symbols represent single gas experiments at 1 bar and 30 °C, open symbols represent 50/50 vol% mixed gas experiments with CO_2 partial pressure of 1 bar).
Parts a–c printed with permission from J. Shen, G. Liu, K. Huang, Z. Chu, W. Jin and N. Xu, *ACS Nano*, 2016, **10**, 3398–3409, Copyright 2016 American Chemical Society. Parts d–f reproduced from ref. 58 with permission from the Royal Society of Chemistry.

Jin and co-workers[56] reported the design and fabrication of a GO/polymer hybrid membrane for gas separation. As shown in Figure 2.9a, polyether block amide (PEBA) was chosen as the polymer matrix because it can form hydrogen bonds between GO nanosheets that enable the GO nanosheets to assemble into several-layered GO stacks within the polymer matrix. TEM characterization shows a homogenous dispersion and an interlayer spacing of these GO stacks subnanometer in size that can exhibit molecular-sieving properties for the CO_2/N_2 gas pair. Positron annihilation spectroscopy

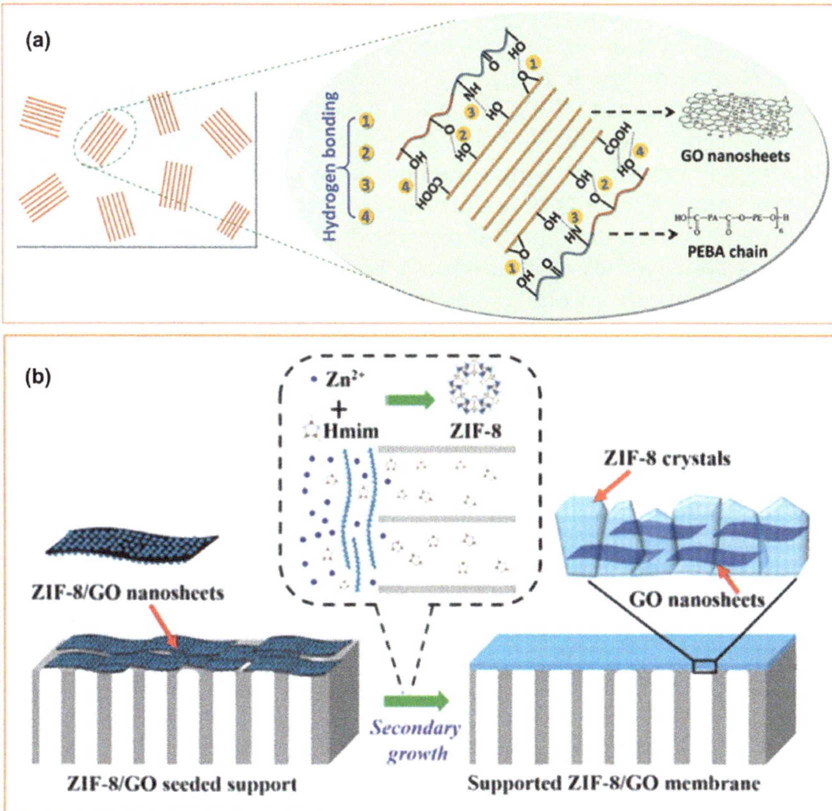

Figure 2.9 Graphene hybrid membranes: (a) schematic representation of the assembly of GO nanosheets in a polymeric environment based on hydrogen bonds formed between different groups on GO and the PEBA chain, (b) illustration of synthesis of ultrathin ZIF-8/GO membrane: coating of flexible ZIF-8/GO nanosheets on a porous support, such as AAO, and subsequent secondary growth by contra-diffusion method.
Reproduced from ref. 6 with permission from John Wiley and Sons, © 2016 Wiley-VCH Verlag GmbH & Co. KGaA, Weinheim.

revealed an increase of free-volumes in the membrane by introducing these GO stacks, providing more transport channels. Meanwhile, sorption measurement indicated that a higher CO_2 sorption capacity and selectivity was obtained in the GO/PEBA hybrid membrane compared to the pristine PEBA membrane. Thus, both the sorption and diffusion of CO_2 over N_2 by the PEBA membrane were enhanced by incorporating GO nanosheets. With fast and selective gas transport channels, only 0.1 wt% GO endowed the GO/PEBA hybrid membrane with an excellent CO_2 separation performance with a CO_2 permeability of 100 Barrer [10^{-10} cm^3 (STP) cm/(cm^2 s cmHg)], CO_2/N_2 selectivity of 91 and operational stability over 100 h. A way to further enhance the gas permeation of this type of hybrid membrane could be constructing GO stacks with vertically aligned interlayer channels in the membrane by controlling the polymer–GO interaction and/or applying an external electro-/magnetic field.

Apart from promoting the transport of water or gas through membranes, GO nanosheets have been widely used as effective nanofillers to prevent biofouling of membranes when applied for water purification. The anti-biofouling property of GO is mainly due to its high surface hydrophilicity, large negative zeta potential, and oxygen-containing groups that inactivate bacteria. A study indicated that adding 1 wt% of GO nanosheets into the polymeric membranes could suppress the microbial adsorption on the membrane surface, thereby resulting in a 5-fold longer time between chemical cleaning for the membrane used for wastewater treatment processes.[61] Similar to the above-discussed pervaporation membranes, the incorporation of GO could improve the hydrophilicity of polymeric membranes for water purification, leading to an enhancement of water flux.

2.4.2 Graphene Oxide/MOF Hybrid Membranes

Besides polymers, graphene materials were hybridized with nanomaterials for membrane separation. Current studies mainly focused on the hybrid of metal–organic frameworks (MOFs) with GO nanosheets. On the one hand, the barrier property of graphene nanosheets was used to seal the defects of crystalline MOF membranes. Huang *et al.*[62] proposed a bicontinuous ZIF-8@GO membrane fabricated through LBL deposition of GO nanosheets on a pre-synthesized semicontinuous ZIF-8 layer. It was demonstrated that the deposited GO layer could seal the non-selective gaps between ZIF-8 crystals by utilizing the capillary forces and covalent bonds. They believed that the gas molecules can only permeate through the micropores of ZIF-8 with a window aperture size of ~0.34 nm, leading to an improved selectivity for separation of H_2 with a molecular kinetic diameter of 0.29 nm from larger gas molecules including CO_2, N_2, CH_4 and C_3H_8. Hu *et al.*[63] employed GO as a 2D platform for growing MOF nanocrystals, and they used the resulting MOF/GO hybrid nanosheets as a seeding layer to synthesize ultrathin MOF molecular-sieving membranes. As shown in Figure 2.9b, the 2D seeding layer acted as a barrier between two different synthesis solutions for the contra-diffusion synthesis process, thereby self-limiting the crystal growth and eliminating defects. The prepared ZIF-8/GO membranes are as thin as ~100 nm, exhibiting molecular-sieving properties for gas separation with a high perm-selectivity of 405 for the H_2/C_3H_8 gas pair.

2.5 Conclusions

This chapter discussed the preparation and application of graphene-based membranes for selectively transporting water, ions and gases. The one-atom-thickness and excellent stiffness suggested porous graphene membranes as ideal molecular sieves, while the precise and large-area perforation of the graphene monolayer remains challenging. Graphene laminate membranes, mostly assembled by GO nanosheets, have been widely studied and shown extraordinary water flow enhancement for water purification and excellent gas separation properties beyond the performance limitation of conventional membranes. Graphene-based hybrid membranes provided a promising alternative for improving the performance of existing membranes.

Future research on graphene-based membranes can be focused on the following aspects: (1) low-cost and controllable synthesis of graphene materials with fewer defects and a specific lateral size; (2) developing approaches to precisely fabricate pore size/interlayer spacing with subnanometer size for efficiently sieving salts from water; (3) understanding the transport mechanism through graphene laminate membranes with delicate pore structures consisting of in-plane structural defects, slit-like pores and interlayer galleries; (4) challenges for practical application in high-pressure, cross-flow or fouling systems.

Acknowledgements

We acknowledge the National Natural Science Foundation of China (Nos. 21406107, 21490585, 21476107, 21776125), National Key Basic Research Program (2017YFB0602500), Innovative Research Team Program by the Ministry of Education of China (No. IRT17R54), Natural Science Foundation of Jiangsu Province (No. BK20140930) and the Topnotch Academic Programs Project of Jiangsu Higher Education Institutions (TAPP) for financial support.

References

1. W. J. Koros and C. Zhang, *Nat. Mater.*, 2017, **16**, 289–297.
2. D. L. Gin and R. D. Noble, *Science*, 2011, **332**, 674–676.
3. H. B. Park, J. Kamcev, L. M. Robeson, M. Elimelech and B. D. Freeman, *Science*, 2017, 356.
4. B. J. Hinds, N. Chopra, T. Rantell, R. Andrews, V. Gavalas and L. G. Bachas, *Science*, 2004, **303**, 62–65.
5. K. S. Novoselov, A. K. Geim, S. V. Morozov, D. Jiang, Y. Zhang, S. V. Dubonos, I. V. Grigorieva and A. A. Firsov, *Science*, 2004, **306**, 666–669.
6. G. Liu, W. Jin and N. Xu, *Angew. Chem., Int. Ed.*, 2016, **55**, 13384–13397.
7. G. Liu, W. Jin and N. Xu, *Chem. Soc. Rev.*, 2015, **44**, 5016–5030.
8. V. Berry, *Carbon*, 2013, **62**, 1–10.
9. J. S. Bunch, S. S. Verbridge, J. S. Alden, A. M. van der Zande, J. M. Parpia, H. G. Craighead and P. L. McEuen, *Nano Lett.*, 2008, **8**, 2458–2462.
10. D. Cohen-Tanugi and J. C. Grossman, *Nano Lett.*, 2012, **12**, 3602–3608.
11. D. Cohen-Tanugi, L.-C. Lin and J. C. Grossman, *Nano Lett.*, 2016, **16**, 1027–1033.

12. K. Sint, B. Wang and P. Král, *J. Am. Chem. Soc.*, 2008, **130**, 16448–16449.
13. D.-e. Jiang, V. R. Cooper and S. Dai, *Nano Lett.*, 2009, **9**, 4019–4024.
14. Z. Tian, S. M. Mahurin, S. Dai and D.-E. Jiang, *Nano Lett.*, 2017, **17**, 1802–1807.
15. L. Wang, M. S. H. Boutilier, P. R. Kidambi, D. Jang, N. G. Hadjiconstantinou and R. Karnik, *Nat Nano*, 2017, **12**, 509–522.
16. L. Huang, M. Zhang, C. Li and G. Shi, *J. Phys. Chem. Lett.*, 2015, **6**, 2806–2815.
17. S. P. Koenig, L. Wang, J. Pellegrino and J. S. Bunch, *Nat. Nano*, 2012, **7**, 728–732.
18. S. P. Surwade, S. N. Smirnov, I. V. Vlassiouk, R. R. Unocic, G. M. Veith, S. Dai and S. M. Mahurin, *Nat. Nano*, 2015, **10**, 459–464.
19. S. C. O'Hern, M. S. H. Boutilier, J.-C. Idrobo, Y. Song, J. Kong, T. Laoui, M. Atieh and R. Karnik, *Nano Lett.*, 2014, **14**, 1234–1241.
20. K. Celebi, J. Buchheim, R. M. Wyss, A. Droudian, P. Gasser, I. Shorubalko, J.-I. Kye, C. Lee and H. G. Park, *Science*, 2014, **344**, 289–292.
21. S. C. O'Hern, D. Jang, S. Bose, J.-C. Idrobo, Y. Song, T. Laoui, J. Kong and R. Karnik, *Nano Lett.*, 2015, **15**, 3254–3260.
22. F. Perreault, A. Fonseca, de Faria and M. Elimelech, *Chem. Soc. Rev.*, 2015, **44**, 5861–5896.
23. D. R. Dreyer, S. Park, C. W. Bielawski and R. S. Ruoff, *Chem. Soc. Rev.*, 2010, **39**, 228–240.
24. D. A. Dikin, S. Stankovich, E. J. Zimney, R. D. Piner, G. H. B. Dommett, G. Evmenenko, S. T. Nguyen and R. S. Ruoff, *Nature*, 2007, **448**, 457–460.
25. Y. Lou, G. Liu, S. Liu, J. Shen and W. Jin, *Appl. Surf. Sci.*, 2014, **307**, 631–637.
26. C.-H. Tsou, Q.-F. An, S.-C. Lo, M. De Guzman, W.-S. Hung, C.-C. Hu, K.-R. Lee and J.-Y. Lai, *J. Membr. Sci.*, 2015, **477**, 93–100.
27. H. Li, Z. Song, X. Zhang, Y. Huang, S. Li, Y. Mao, H. J. Ploehn, Y. Bao and M. Yu, *Science*, 2013, **342**, 95–98.
28. P. Sun, M. Zhu, K. Wang, M. Zhong, J. Wei, D. Wu, Z. Xu and H. Zhu, *ACS Nano*, 2013, **7**, 428–437.
29. H. W. Kim, H. W. Yoon, S.-M. Yoon, B. M. Yoo, B. K. Ahn, Y. H. Cho, H. J. Shin, H. Yang, U. Paik, S. Kwon, J.-Y. Choi and H. B. Park, *Science*, 2013, **342**, 91–95.
30. R. R. Nair, H. A. Wu, P. N. Jayaram, I. V. Grigorieva and A. K. Geim, *Science*, 2012, **335**, 442–444.
31. K. Guan, J. Shen, G. Liu, J. Zhao, H. Zhou and W. Jin, *Sep. Purif. Technol.*, 2017, **174**, 126–135.
32. M. Hu and B. Mi, *Environ. Sci. Technol.*, 2013, **47**, 3715–3723.
33. W.-S. Hung, C.-H. Tsou, M. De Guzman, Q.-F. An, Y.-L. Liu, Y.-M. Zhang, C.-C. Hu, K.-R. Lee and J.-Y. Lai, *Chem. Mater.*, 2014, **26**, 2983–2990.
34. A. Akbari, P. Sheath, S. T. Martin, D. B. Shinde, M. Shaibani, P. C. Banerjee, R. Tkacz, D. Bhattacharyya and M. Majumder, *Nat. Comm.*, 2016, **7**, 10891.

35. M. Zhang, K. Guan, J. Shen, G. Liu, Y. Fan and W. Jin, *AIChE J.*, 2017, **63**, 5054–5063.
36. Y. Han, Z. Xu and C. Gao, *Adv. Funct. Mater.*, 2013, **23**, 3693–3700.
37. L. Qiu, X. Zhang, W. Yang, Y. Wang, G. P. Simon and D. Li, *Chem. Commun.*, 2011, **47**, 5810–5812.
38. R. K. Joshi, P. Carbone, F. C. Wang, V. G. Kravets, Y. Su, I. V. Grigorieva, H. A. Wu, A. K. Geim and R. R. Nair, *Science*, 2014, **343**, 752–754.
39. L. Huang, Y. Li, Q. Zhou, W. Yuan and G. Shi, *Adv. Mater.*, 2015, **27**, 3797–3802.
40. S. Zheng, Q. Tu, J. J. Urban, S. Li and B. Mi, *ACS Nano*, 2017, **11**, 6440–6450.
41. Y. Su, V. G. Kravets, S. L. Wong, J. Waters, A. K. Geim and R. R. Nair, *Nat. Commun.*, 2014, **5**.
42. X. Chen, M. Qiu, H. Ding, K. Fu and Y. Fan, *Nanoscale*, 2016, **8**, 5696–5705.
43. D. He, L. Shen, X. Zhang, Y. Wang, N. Bao and H. H. Kung, *AIChE J.*, 2014, **60**, 2757–2764.
44. W. Wang, E. Eftekhari, G. Zhu, X. Zhang, Z. Yan and Q. Li, *Chem. Commun.*, 2014, **50**, 13089–13092.
45. S. J. Gao, H. Qin, P. Liu and J. Jin, *J. Mater. Chem. A*, 2015, **3**, 6649–6654.
46. Y. Han, Y. Jiang and C. Gao, *ACS Appl. Mater. Interfaces*, 2015, 7, 8147–8155.
47. K. Guan, D. Zhao, M. Zhang, J. Shen, G. Zhou, G. Liu and W. Jin, *J. Membr. Sci.*, 2017, **542**, 41–51.
48. H. Huang, Z. Song, N. Wei, L. Shi, Y. Mao, Y. Ying, L. Sun, Z. Xu and X. Peng, *Nat. Commun.*, 2013, **4**.
49. L. Ding, Y. Wei, Y. Wang, H. Chen, J. Caro and H. Wang, *Angew. Chem., Int. Ed.*, 2017, **56**, 1825–1829.
50. J. Shen, G. Liu, K. Huang, Z. Chu, W. Jin and N. Xu, *ACS Nano*, 2016, **10**, 3398–3409.
51. J. Abraham, K. S. Vasu, C. D. Williams, K. Gopinadhan, Y. Su, C. T. Cherian, J. Dix, E. Prestat, S. J. Haigh, I. V. Grigorieva, P. Carbone, A. K. Geim and R. R. Nair, *Nat. Nano*, 2017, **12**, 546–550.
52. K. Huang, G. Liu, Y. Lou, Z. Dong, J. Shen and W. Jin, *Angew. Chem., Int. Ed.*, 2014, **53**, 6929–6932.
53. K. Huang, G. Liu, J. Shen, Z. Chu, H. Zhou, X. Gu, W. Jin and N. Xu, *Adv. Funct. Mater.*, 2015, **25**, 5809–5815.
54. K. Goh, W. Jiang, H. E. Karahan, S. Zhai, L. Wei, D. Yu, A. G. Fane, R. Wang and Y. Chen, *Adv. Funct. Mater.*, 2015, **25**, 7348–7359.
55. P. Sun, F. Zheng, M. Zhu, Z. Song, K. Wang, M. Zhong, D. Wu, R. B. Little, Z. Xu and H. Zhu, *ACS Nano*, 2014, **8**, 850–859.
56. J. Shen, G. Liu, K. Huang, W. Jin, K.-R. Lee and N. Xu, *Angew. Chem., Int. Ed.*, 2015, **54**, 578–582.
57. S. Wang, Y. Wu, N. Zhang, G. He, Q. Xin, X. Wu, H. Wu, X. Cao, M. D. Guiver and Z. Jiang, *Energy Environ. Sci.*, 2016, **9**, 3107–3112.

58. K. Cao, Z. Jiang, J. Zhao, C. Zhao, C. Gao, F. Pan, B. Wang, X. Cao and J. Yang, *J. Membr. Sci.*, 2014, **469**, 272–283.
59. B. G. Choi, J. Hong, Y. C. Park, D. H. Jung, W. H. Hong, P. T. Hammond and H. Park, *ACS Nano*, 2011, **5**, 5167–5174.
60. H. Zarrin, D. Higgins, Y. Jun, Z. Chen and M. Fowler, *J. Phys. Chem. C*, 2011, **115**, 20774–20781.
61. J. Lee, H.-R. Chae, Y. J. Won, K. Lee, C.-H. Lee, H. H. Lee, I.-C. Kim and J.-M. Lee, *J. Membr. Sci.*, 2013, **448**, 223–230.
62. A. Huang, Q. Liu, N. Wang, Y. Zhu and J. Caro, *J. Am. Chem. Soc.*, 2014, **136**, 14686–14689.
63. Y. Hu, J. Wei, Y. Liang, H. Zhang, X. Zhang, W. Shen and H. Wang, *Angew. Chem., Int. Ed.*, 2016, **55**, 2048–2052.
64. W.-S. Hung, Q.-F. An, M. De Guzman, H.-Y. Lin, S.-H. Huang, W.-R. Liu, C.-C. Hu, K.-R. Lee and J.-Y. Lai, *Carbon*, 2014, **68**, 670–677.
65. G. Li, L. Shi, G. Zeng, Y. Zhang and Y. Sun, *RSC Adv.*, 2014, **4**, 52012–52015.
66. T.-M. Yeh, Z. Wang, D. Mahajan, B. S. Hsiao and B. Chu, *J. Mater. Chem. A*, 2013, **1**, 12998.
67. X. Chen, G. Liu, H. Zhang and Y. Fan, *Chin. J. Chem. Eng.*, 2015, **23**, 1102–1109.
68. K. Huang, J. Yuan, G. Shen, G. Liu and W. Jin, *Chin. J. Chem. Eng.*, 2017, **25**, 752–759.
69. C. Chi, X. Wang, Y. Peng, Y. Qian, Z. Hu, J. Dong and D. Zhao, *Chem. Mater.*, 2016, **28**, 2921–2927.

CHAPTER 3

Mass Transport Across Atomically Thin Membranes

K.-P. SCHLICHTING AND H. G. PARK*

Nanoscience for Energy Technology and Sustainability, Department of Mechanical and Process Engineering, Eidgenössische Technische Hochschule (ETH) Zürich, Tannenstrasse 3, Zürich CH-8092, Switzerland
*Email: parkh@ethz.ch

3.1 Introduction

Membrane separation technology is drawing growing attention today in research and technology due to its promise of process intensification toward resource saving and process simplicity.[1] The simplicity stems from the nature of a separation membrane, a barrier that allows disproportionate transport of one species over the other. Chemical separation results from this transport disproportionation. Membrane-based separation can consume far less energy and resources than other separation technologies do based on an energy intensive phase change (*e.g.* cryogenic distillation in gas separation or thermal distillation in water treatment).[2,3] Despite this promise, the widespread economic use of membranes in separation applications has not been seen yet, which is ascribable to limits in material properties, challenges of upscaling, and issues of economic fabrication. One of the overarching research goals of membrane-based separation technology is the establishment of enhanced membrane performance to exploit the technological and economic potential and to promote the engineering efforts toward membrane upscaling and cost-effective process optimization.

Nanoscience & Nanotechnology Series No. 47
Graphene-based Membranes for Mass Transport Applications
Edited by Pengzhan Sun and Hongwei Zhu

Published by the Royal Society of Chemistry, www.rsc.org

Separation membranes can be largely classified into two categories: dense membranes, *i.e.*, membranes not possessing definable porous structures, whose transport mechanism is conveniently described by the solution-diffusion model; and porous membranes, for which the hydrodynamic and sub-continuum transport models are often utilized.[2,4] To describe the separation performance of a membrane, its permeance – the transport rate per unit area normalized by a driving force – and selectivity – the permeance ratio of the species of interest – comprise the main figures of merit. In general, the permeance of a membrane material is inversely proportional to its thickness, since friction forces with pore walls or diffusion pathways decrease by shortening the effective transport lengths. This inverse relationship leads to the desire of making ever thinner membranes to enhance the performance. Selectivity of dense membranes, on the other hand, depends typically on the solubility and diffusion ratios of the permeating species and is therefore an intrinsic property of the membrane material. Contrary to dense membranes, porous membranes separate on the basis of molecule–pore interactions, related to the ratio between pore size and molecular length scale. This interaction of porous membranes allows the selective transport behaviour in principle to be decoupled from the intrinsic material properties, potentially offering flexibility in the determination of their selectivity. The thickness of porous membranes contributes to the permeance by controlling the entire amount of the molecule–pore friction mechanism.

For a given pore size, the membrane thickness can be decreased all the way down to an atomic dimension that graphene and 2D materials can offer. At the atomic dimension of membrane thickness, it is expected that the friction and viscous interaction of transporting fluid in the pore interior disappear or are minimized to permit very rapid permeation. In fact, in the limit of vanishing thickness, the transport pathway is supposed to cause a singularity in the transport rate based on continuum model prediction. However, as shall be discussed later in this text, pores formed on graphene can help avoid the Fickian singularity in a way that omission of conventional determinants of transport physics brings about the emergence of other transport-governing phenomena related to the 3D-to-2D transition. Ultimately, in the limit of the 2D material thickness, the lattice spacing of 2D crystals can be considered similar to the free volume in dense membranes such that it raises a question of whether a description of 2D membranes by a generalized solution diffusion mechanism may work out. For this reason, it is important to study the transport physics across 2D membranes because conventional models may not be able to accurately describe mass transport across graphene membranes.

A prerequisite question to answer prior to the utilization of graphene as a porous membrane is whether it is permeable in its pristine form to the species it should separate. In 2008, Bunch and co-workers performed experiments to answer this question.[5] They pressurized a micrometre-sized cavity in SiO_2 covered by mechanically exfoliated graphene flakes and subsequently measured the deflection of the bulging graphene with an atomic force microscope. For various gases including He, they could show that the deflection over time of the pressurized graphene blister is independent of the layers of graphene (up to 75 layers were tested) and simultaneously matches the diffusion rate through their substrate. Further analysis of a tunnelling probability across pristine graphene revealed that pristine graphene is impermeable to gases as small as He and may pose a formidable barrier to mass transport. Consequently, in its pristine form, graphene is not directly suited as a separation membrane unless pores are introduced into its structure. Pore formation has been immediately demonstrated experimentally *via* energetic electron beam irradiation,[6] energetic ion beam bombardment,[7] heterogeneous oxidation through plasma[8] or harsh chemistry.[9] Notably, the pores created in these ways remain stable for an intermediate-to-long period of time, shedding light on the feasibility of porous graphene for membrane applications.

3.2 Gas Transport Across Porous Graphene Membranes

The study of gas transport across porous graphene membranes has been motivated by graphene's potential impermeability to gases despite the atomic thinness. Once the graphene is chopped open to bear tiny pores, the thinness of graphene in turn acts as an ultimate gas permeator. In this light, the focus of gas transport study across porous graphene has been geared toward engineering highly selective membrane materials. High selectivity is expected for pores that are of similar size to the kinetic diameters of the gas molecules to be separated, a mechanism known as molecular sieving.

The pioneers of gas separation applications of porous graphene are Jiang and co-workers, who have used density functional theory (DFT) to investigate the transport behaviour of light inorganic gases through subnanometre pores on graphene.[10] After removing two hexagonal carbon rings (10 carbon atoms) from a graphene lattice in a modelling domain, they considered two types of pores: one by hydrogen passivation of all the carbon dangling bonds, and the other by nitrogen and hydrogen passivation. Such passivation constricted the pore width from 3.0 Å to 2.5 Å on the basis of the isoelectric surface of the electron density distribution of the pore edge atoms (0.02 e/Å^3). Ballistic transport of H_2 and CH_4 across these model pores undergoes diffusion barriers describable with activation energies of an Arrhenius formula (assuming an identical exponential prefactor). Once determining the smallest width of a pore to dictate the permeance, they found out that molecules orient preferentially against the pore edges during the passage and that this orientation is reliant on the molecule and edge-passivation pair. The H_2/CH_4 selectivity has been reported to be as high as 10^8 for the 3.0 Å pore and 10^{23} for the smaller pore. They attributed the 15-order-of-magnitude rise in the selectivity to the exponential dependence of the gas diffusion barrier, elevating greatly when the constricting width becomes commensurable with the kinetic diameters of the transporting molecules (2.89 Å for H_2 and 3.80 Å for CH_4). Even though they have not extended the modelling beyond the ballistic transport for simplicity, this study has predicted the potentially astounding selectivity of graphene membranes for gas separation and proposed an activation energy of pore passage as an indicator of the cross-graphene-pore gas transport.

Following this pioneering work, Li *et al.* investigated the permeation of H_2, CO, CO_2, and CH_4 with DFT by considering a graphene pore as large as one benzene-ring size passivated by hydrogen atoms.[11] They show an increasing passage energy barrier for increasing isoelectric surface overlap of molecule and pore in the transition state, elucidating that the electron density distributions of pore and molecule are rooted in the repulsion. Extending similar analysis to noble gases such as Ne, He, and Ar, Blankenburg *et al.* reported that noble gases follow a pathway of a minimum potential energy valley, attributed to attractive electrostatic interaction caused by the large polarizability of the noble gases.[12] NH_3 can also permeate following the minimum potential pathway due to its capability of hydrogen-bond formation with the hydrogen-terminated graphene pore edge. Also, the membrane can deform during the passage of gases due to their strong interaction, which lowers the energy barrier effectively compared with rigid pores. These findings support the fact that not only size comparison but also the various physical and chemical interactions between the graphene pore and the passing molecules can participate in the cross-graphene-pore gas transport mechanism. The pore deformation phenomenon has been subsequently investigated in more detail by Hauser and Schwerdtfeger for gases crucial in natural gas processing.[13] With a graphene pore system similar to Jiang and colleagues' pores, they observed that the pores adapt their size after removal of the benzene rings and they also monitored strong pore deformation during the passage of various gas species. For the nitrogen-terminated pore, they also observed a slight bending of the CO_2 molecule during the passage, explainable by a Lewis-acid–base mechanism of the charge distribution of bent CO_2 that shows a propensity for the Lewis-base nitrogen. The non-negligible interaction between transporting molecule and pore edge can cause molecular deformation during gas permeation. Thus, a simple comparison between molecular and pore sizes loses its significance beyond a first approximation of permeation/separation.

The first molecular dynamics (MD) simulation of gas transport (H_2 and N_2) across the graphene pore was carried out by Du *et al.*, scanning the pore sizes from 10 C equivalence to 32.[14] For the smallest pores, the pore size is smaller than the kinetic diameter of N_2, and so they observed no N_2 passage. Unexpectedly though, as pores opened up large enough to pass N_2, the calculated N_2 flow rate exhibited a superlinear proportionality with the open pore area, whereas H_2 followed the pore area scaling quite linearly. From the MD simulation, they found that a layer of N_2 molecules can absorb onto the graphene surface *via* van-der-Waals interactions, resulting in a surface diffusion of the molecules that outpaces the ballistic diffusion. The unveiled surface adsorption and diffusion of gas molecules at pore dimensions slightly larger than the molecular sieving regime are of significant practical importance; for example, the surface diffusion mechanism could deteriorate an otherwise molecular-sieving-level separation factor or could enhance the selectivity toward strongly adsorbing species. Preferential adsorption onto the graphene surface has also been confirmed. According to an investigation of adsorption of H_2, CO_2, CH_4, and N_2, charge variation of the membrane surface has an influence on the physisorption of the gases.[15] A subsequent DFT calculation for hydrogenated pores with the size of two benzene rings added that the large quadrupole moment of CO_2 can be related to stronger adsorption to the graphene surface compared to CH_4, N_2, and O_2.[16] Further analyses by MD simulations found that linear molecules can absorb flat onto the surface and that CO_2 requires several events for the complete passage of the pore; for the passage, the adsorbate should be oriented properly to overcome the barrier, an event unlikely to happen to molecules approaching from the gas phase to the pore directly. These findings were simplified to a Langmuir adsorption model, leading to an argument that the surface-diffusion-related pore passage is the single, rate-limiting step. Competitive adsorption had a negligible impact on the selectivity because the gas adsorption did not fully saturate the membrane surface in the temperature and pressure ranges scanned in this investigation.

On the other hand, Drahushuk *et al.*[17] proposed a five-step analytical model by making an analogy of the surface-diffusion-incorporated transport to a surface-site-mediated catalytic reaction. The five identified steps are surface adsorption (on the feed side), pore association, pore passage (engagement), surface diffusion (on the permeate side), and surface desorption (detachment). Steady-state solutions of the differential equations of the analytical model for a range of graphene pore sizes from the 2-benzene-ring equivalent to the 32-carbon-atom equivalent suggested that "pore passage" is the rate-limiting step when the pore size is close to that of molecular sieves whereas "surface diffusion" can become the rate-limiting step for larger pores. They further predicted the importance of pressure-dependent permeance and the influence of feed composition due to competitive adsorption, in contrast to direct gas-phase passage. Sun *et al.* further compared the relative contributions of direct and surface diffusions for He, H_2, N_2, and CH_4 by MD simulation as a function of pore size.[18] The surface diffusion mechanism may contribute significantly to the entire permeation, such that its contribution could add up to being 16 times higher than the direct transmission part for transport of such a strongly absorbing species as CH_4 across a 12-C-removal equivalent pore. Surface diffusion may still take part in the *trans*-graphene transport, by at least an equal amount of contribution, for weakly interacting gases like He and H_2.

The effect of pore functional groups on permeation, a topic addressed briefly early on in the field, has been given renewed attention. Shan *et al.* looked into CO_2/N_2 separation across pores of 10–19 C atom equivalence by

MD simulations.[19] Pore functionalization with nitrogen could render the CO_2/N_2 selectivity as high as 11 for the otherwise unselective pores, attributable to a higher electrostatic interaction between the nitrogen edge and CO_2 rather than with $N_{2,}$, since CO_2 has a higher quadrupole moment. They also showed that functionalization of the membrane surface with hydroxyl groups could lead to preferential adsorption of CO_2 over N_2 by about seven times. These two findings agree with the previous observation that the selectivity toward CO_2 relies on the feed partial pressure of CO_2. Strong interaction among CO_2 molecules that can help to overcome the passage barrier turns vigorous with increased partial pressures.

The significance of the preferential molecular orientation in the *trans*-graphene passage noted earlier was confirmed by MD simulations. Solvik *et al.* examined the separation of various olefins and paraffins using all-hydrogenated pores of two-hexagon equivalence.[10,12,16,20] They considered in their simulation the free deformation of the membrane and its pores that had been recognized as crucial in the analysis yet often neglected. The simulation showed surface adsorption limited transport with preferential permeation, or separation, of ethene over ethane. Interestingly, the ethane, the molecule that adsorbs more, is permeating less compared to the less adsorbing molecule ethene. They explain their results by an unfavourable entropic energy barrier for ethane during passage. During the *trans*-graphene passage, molecules fall into a potential well, though ethane does not reside well inside the pore compared with ethene. The researchers decomposed the total barrier of the permeation rate model into activation enthalpy, surface adsorption, and entropic terms and observed that the activation enthalpy is similar, so that surface adsorption would in fact favour passage of ethane. However, the entropic barrier causes the ethene transport prevalence across the membrane. This finding is rationalized by the small size of the ethane molecule such that more configurations within the pore are possible during passage. Hence, it's entropic penalty is lower. Another MD simulation looking into the molecular orientation during the passage of CO_2 and H_2S across 2–3-benzene-ring-equivalent, H- or N-functionalized pores found that the linear CO_2 molecule follows a narrow range of orientation during passage.[21] The radial density distribution extending out of the pore exhibits an accumulation zone within the pores, surrounded by an ~0.5-nm-wide depletion zone, which eventually transitions into a peripheral zone of constant adsorption density. Starting from the adsorption zone, the molecules feel no concentration gradient and thus accumulate homogeneously. Near the pore, on the other hand, a concentration or density gradient develops in the pore, along which molecules diffuse.

Recently, another aspect of permeation across subnanometre openings has been predicted by combined DFT and MD simulations of CH_4, CO_2, and N_2 separation.[22] The barrier against the molecular passage consists of a combination of pore entry and exit barriers and varies depending on the pore, functionalization, and the passing molecule. For example, CO_2 usually occupies a potential well within the pore. To leave the pore, the molecule must escape the potential energy barrier. Once charge is imposed at the pore edge, its strong interaction with the CO_2 quadrupole can adjust the transport energy barrier against CO_2. This mechanism provides an additional means of separation other than size exclusion or hindered diffusion and may be particularly relevant for similarly sized molecules.

It is interesting to consider the effect of a non-permeating species in a gas mixture on the permeance of a permeating mixture. Wen *et al.*'s MD simulation delved into this question using CH_4 as a non-permeating species and H_2 and N_2 as permeating species through a 13-C-atom-missing pore.[23] The presence of CH_4 can decrease both permeances of H_2 and N_2, and this effect can be more severe for N_2. They attribute this to two contributions: competitive adsorption and pore blocking. Competitive adsorption is more severe for N_2 as H_2 barely adsorbs to the surface. N_2, on the other hand, tends to adsorb significantly and puts itself into competition with CH_4, resulting in less surface occupancy than in the absence of CH_4. Besides, the number densities of both H_2 and N_2 can increase near the pore, though the permeation is curbed by the presence of CH_4. According to their explanation, CH_4 may partly occupy the subnanometre pores and slow down the transport of H_2 and N_2 effectively. The combined effect of the competitive adsorption and the pore blocking can result in decreased permeance in the presence of a non-permeating species, implying practical importance in gas separations in which more than two species are often present.

Overall, the available literature revealed various facets of the permeation mechanism of gaseous molecules across porous graphene membranes. Besides an energy barrier resulting from an interaction between a subnanometric pore and transporting molecules, in association with electron cloud overlapping, adsorption of the molecules to the graphene surface and subsequent surface diffusion have a significant influence on the permeation and selectivity of the graphene membranes with subnanometric pores. Furthermore, the effects of the pore and surface functionalization, an entropic barrier based on orientation restriction, as well as the mixture effects such as competitive adsorption and steric hindrance (*i.e.*, pore blocking) can altogether play a non-negligible role in *trans*-graphene membrane transport. It is noteworthy that various simulations predict a wide range of permeance and selectivity values, calling for verification to clarify the deterministic transport mechanisms and to ultimately quantify membrane performance for gas separation.

In this light, the experimental understanding of the mass transport across porous graphene membranes can fall into two categories. The first category comprises publications based on mechanically exfoliated graphene flakes that are high in quality but only as wide as tens of micrometres. With these systems, it was possible to study impermeability of and transport across subnanometric pores on graphene in the molecular sieving regime. The second category covers publications based on the graphene grown by chemical vapour deposition (CVD) and incorporating intrinsic or artificial pore formation. CVD can synthesize potentially large-scale membranes and is considered the most promising manufacturing method for real graphene membrane applications. The large area allows for various manufacturing options for the graphene perforation and the membrane formation. Intrinsic defects and ruptures of the synthesized graphene could be a drawback of this approach during membrane manufacturing, but they can provide a molecular transport pathway per se.

The prerequisite for studying the transport across porous graphene is the impermeability of the pristine graphene demonstrated by Bunch and colleagues.[5] Employing the same method, Koenig *et al.* pressurized mostly bilayer graphene blisters by applying UV-oxidation etching.[24] If well controlled, this treatment may enable the formation of pores in the subnanometre dimension, a characteristic size regime useful for molecular sieving of various gases. The transport rates were either measured by AFM tracking of the membrane deflection over time or by a temporal change in resonance frequency of the membrane while transporting gases. They reported giant selectivity between 10^3 (for H_2/Ar) and 10^4 (for H_2/N_2) of the UV-oxidated bilayer graphene membrane as well as size exclusion of SF_6 having a kinetic diameter of 4.9 Å. The measured permeance matches the prediction of Blankenburg *et al.* of a H-passivated, 6-C-atom-equivalent pore, but interestingly, their selectivity of 2 for H_2/CO_2 differs significantly from the theoretical prediction (10^{17}).

Further work of the same group with a monolayer graphene blister having subnanometric pores has revealed that He permeance can vary by a factor of ~5, if the membrane surface is irradiated with a laser of 2–2.3 eV photon energy.[25] Additional laser shining could return the permeance to its initial value. They attributed this switchable permeance to the presence of gold nanoparticles that had been evaporated onto the surface. According to their speculation, energy input such as photon irradiation could induce surface migration of Au NPs to block the subnanometric pores. However, what remains to be clarified is a negative control of this permeance switching phenomenon for a graphene membrane without the nanoparticles on it.

Actually, this paper reported stochastic Ne permeance switching of monolayer graphene in the absence of the surface nanoparticles. Their subsequent work provides the Hidden-Markov-model-based analysis of this stochastic permeance switching of three monolayer graphene pores that can switch their individual permeance values between "high" and "low" states.[26] For gas species having kinetic diameters larger than that of He, the permeance shows various discrete levels that can be mathematically modelled by this binary state system of the three pores. From the switching frequency, the authors could estimate the activation energy value of 1 eV required to switch the permeance state, consistent with the energy barrier of *cis–trans* isomerization rearrangements. Interestingly, this energy is less than the laser photon energy previously reported to switch a molecular valve from "high" to "low".[25]

Due to the size limitations of graphene flakes, mechanical exfoliation has not been considered realistic for membrane applications outside a laboratory. Instead, CVD-grown graphene can potentially take as large a dimension as a few metres such that its use as a large-scale gas separation membrane is feasible. Boutilier *et al.* investigated the feasibility of achieving gas selective graphene membranes despite the presence of non-selective defects, ruptures, and incomplete graphene coverage of the support structure.[27] They could show an exponential decrease of leakage pathways across CVD-grown graphene as well as increased gas permselectivity for multiple layers of graphene, confirming that both the total amount of leakage as well as the average size of the leakage path can diminish if multiple layers of imperfect CVD-grown graphene are stacked atop one another. In order to predict the selectivity of graphene, a model was proposed considering a membrane that possesses hypothetical, highly selective subnanometric pores and lies on a porous support, a membrane architecture so-called a thin-film composite. This model illustrates that, despite the presence of unselective defects in the graphene, it is possible to obtain a highly selective membrane as long as the contribution of the unwanted defects is negligibly small. The authors argue that the support should slow down the non-selective transport through ruptures and only slightly reduce overall selectivity.

Later, they used the comparable transport impedances of graphene and the support layer to show enhanced Knudsen diffusion permselectivity of an ion bombarded and subsequently PDMS spin-coated PDMS/graphene/PCTE composite.[28] A further study used trilayer graphene on 20-nm-pored anodic aluminium oxide (AAO) that was etched in O_2 plasma for various durations. Single-gas permeance measurements revealed a permselectivity above Knudsen, indicative of molecular sieving.[29] However, the contribution of unselective broken graphene with potentially highly selective subnanometric pores in graphene hampers the unambiguous characterization of the graphene pores. Therefore, a model was utilized to estimate the selectivity of graphene pores in the subnanometric regime, when accounting for effects of ruptures and imperfect coverage of the graphene layer. The model predicts He/SF_6 selectivity ranging from 10 to 400 depending on the fabrication conditions. The permeance of their measured composite membranes is quite small due to the small pores of the support membrane such that it is comparable to permeances of <100-nm-thick polymeric membranes. Consequently, these composite structures cannot meet the promise of ultimate permeation of atomically thin membranes. These studies highlight the significant challenges of gas separation using porous graphene on macroscopic areas and suggest strategies to overcome these difficulties.

As the thickness of a pore approaches zero, transport physics will drop out the channel-length dependency along with one pore-width dimension, thus avoiding the singularity – an infinite mass flux – of the Fickian dynamics. The transport dynamics enters into non-Fickian physics. In a continuum mechanics regime, where this zero-thickness pore measures an opening size far in excess of the molecular interaction length scales, a Stokes flow model can provide analytical solutions. For a pressurized fluid flow across an opening through a 2D disc, for instance, the linear relation between the flux and the pressure forcing has been analytically obtained in the form of Sampson's formula.[30,31] In the Sampsonian dynamics, the mass flux of a fluid is linearly proportional to the pore opening size (diameter) and the pressure force, inversely proportional to the fluid viscosity, and invariant to the pore length (or depth). On the other hand, in a molecular flow regime, where pore size is comparable to or far smaller than the molecular interaction length scales, a ballistic transport model such as effusion or the dusty gas model can describe the transport dynamics. Effusion can occur for a gas species when both the pore size and the depth are much smaller than the mean free path of the species. The simplest description of effusion is related to the Boltzmann gas dynamics; the mass flux across an aperture on a thin wall is proportional to the disparity in the product of gas density and thermal velocity. As a result, the effusive transport flux from one side to the other depends linearly on the pressure and is inversely proportional to the square roots of molar mass and temperature. Inherent to the nature of the free molecular transport, the flux from one side to the other is independent of the counter flux, and thus the net flux is determined by the difference in both types of flux.

It has been difficult to verify these flow theories and characterize them experimentally because of the deficiency of a method for manufacturing a nearly zero-thick pore. Characterization of the ballistic effusion could have been made in the high vacuum setup, yet this capability is still bounded to the Knudsen diffusion limit if pore depth cannot be made ultrathin. One fundamental limit of the material thickness is an atom, and it is graphene, one of the ideal 2D materials, that has the mechanical, physical and chemical potential to demonstrate zero-thickness pores. Still, the primary difficulty in the verification of the aforementioned flow theories and the characterization of the nature of the length-invariant transport physics lies in the method of perforating the variously sized pores accurately on the graphene in its unsupported, self-sustaining configuration.

Researchers have recently developed a physical perforation method for freestanding graphene to investigate this length-invariant mass transport across graphene pores in a wide range of transport regimes from free molecular to continuum. Celebi, Buchheim *et al.* applied the focussed ion beam (FIB) technique to drill well-defined pores with diameters ranging from sub-10 nm to 1000 nm on freestanding double-layer graphene (as thick as approximately two C atoms).[7] Preparation of a myriad of similarly sized pores can facilitate the direct atmospheric characterization of the gas transport rate across, for example, pores smaller than the gas mean free paths. With this technical breakthrough, they verified that the effusion mechanism dominates transport in the free molecular regime, whereas in the continuum flow regime of larger pores, a modified Sampson's formula describes the non-Fickian dynamics. Furthermore, a transport minimum at a Knudsen number around unity, often observed for finitely thick channels and ascribable to gas-and-pore-wall friction, was unobserved for the graphene pore, confirming the unique transport mechanism across the 2D pore. Gas permselectivity is scaled to the inverse square root of molar mass just as predicted by Graham's law of effusion.

Despite the theoretical and experimental investigations that have revealed many findings on the gas separation behaviour and the transport mechanism across porous graphene membranes of various pore sizes, important questions still remain unanswered. Particularly for graphene membranes with few-nm to sub-nm pores, a unified picture of selectivity and permeance remains to be established considering the interplay of molecular-level interactions among the membrane, permeating gas, and other gas species. The effect of surface diffusion may cause a deviation from ever larger selectivity for ever smaller pores, as predicted for H_2–N_2 separation for sub-nanometric pores. Figure 3.1 shows a comparative analysis of the H_2–N_2 separation factor for pores of various sizes and sources. The separation

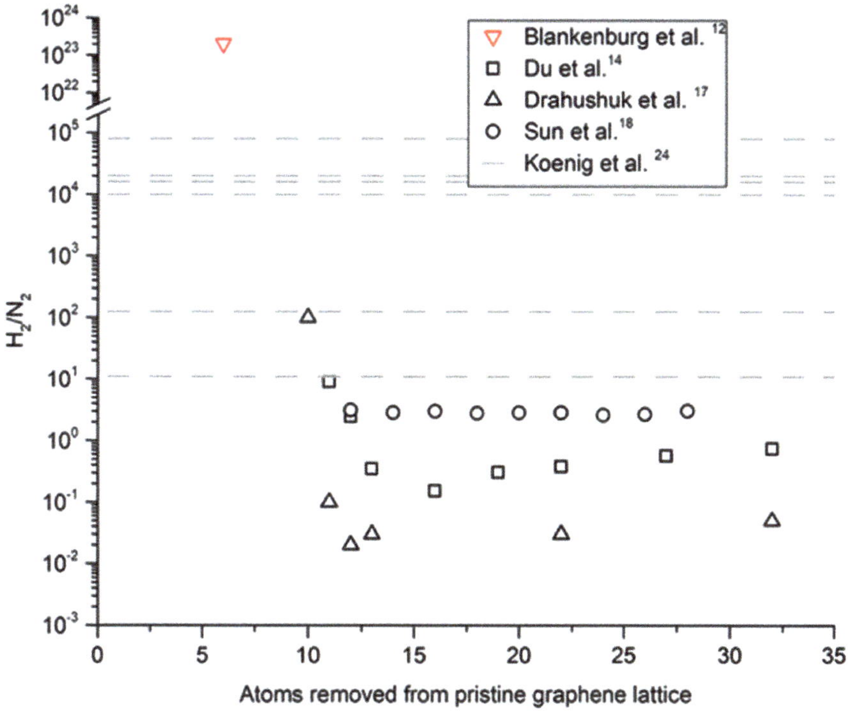

Figure 3.1 H_2–N_2 selectivity values from various simulations and experiment, with respect to graphene pore size (number of C atoms removed from a graphene crystal). A strong disagreement exists among simulations for pores with size equivalence of more than ten C atoms. Symbols represent the corresponding study; colours represent chemical group at the pore edge. Black and red represent no and hydrogen functionalization, respectively. Grey represents unknown functionalization.

factor predictions vary rather largely by up to two orders of magnitude and also conflict with one another with respect to the transport favouring species: *i.e.*, hydrogen-selective or nitrogen-selective. This uncertainty needs clarification about the true separation factor of pores near 1 nanometre in size. Furthermore, the membrane-based gas separation application demands the actual molecular sieving selectivity of CVD-grown defective graphene in the practical mixture separation process. Following this reasoning, the challenge of scaling up the membrane area needs to be taken up, as practical membranes will require membrane areas ranging from metres to even thousands of metres, although this argument remains to be further discussed in the community. For subnanometric pores, the self-healing phenomenon can pose a great challenge to the membrane lifetime.[32] Here, effective approaches for pore edge stabilization by functionalization or other means need to be established.[33,34] Eventually, porous graphene membranes need to compete with other types of gas separation membranes such as zeolites, carbon molecular sieves, carbon nanotube membranes, graphene oxide, metal organic frameworks, and others.[1] Chemical stability, mechanical strength, temperature and pressure requirements, as well as an appropriate support material, need to be investigated to find the best suited material for a targeted gas separation application.

3.3 Liquid Transport Across Porous Graphene Membranes

Liquid transport across porous graphene membranes is of interest not only for the fundamental understanding of transport phenomena but also for various applications potentially encompassing from filtration and desalination to biomedical engineering processes such as dialysis and DNA sequencing.

3.3.1 Water Transport Across Porous Graphene Membranes

The transport of water across graphene membranes was investigated in Suk and Aluru's MD simulation that considered 0.75-nm-wide and 2.75-nm-wide pores on graphene, and the results were compared to the state-of-the-art fast water transport through carbon nanotubes (CNT).[36,37] A very small (~0.8 nm) pore of graphene can transport water at a slightly slower rate than that of a similarly wide CNT conduit, which is related to the frequency of water dipole rearrangement during the passage. For both graphene- and CNT pores, water molecules take a single-file configuration in transporting in these tiny pores. However, water molecules in the 0.8-nm-wide CNT keep a single dipole orientation most of the time during the passage, while water dipoles in the graphene pore can frequently alternate in a rather random orientation. This fluctuation in the dipole orientation requires energy, and thus transport in the subnanometric graphene pores can occur at a slower rate than the water conduction across a CNT with a similar diameter. Across 2.75-nm-wide pores, on the contrary, the simulation resulted in water transport rates faster for graphene pores than CNTs. Unlike a pluglike profile in the CNT, the velocity profile of the water flow demonstrates a reportedly[37] parabolic profile in the graphene pores. A follow-up simulation investigated water transport across graphene pores with various diameters from subnanometre to ∼4 nm diameter and tried to see if it can be modelled with a continuum dynamics theory: an adjusted Hagen–Poiseuille (HP) equation.[35] Instead of the membrane length (0.54 nm), a pore-diameter-dependent hydrodynamic membrane length was utilized with accounting for entry and exit pressure losses, which occur before and after the pore. Regarding the total flow enhancement, the authors observed an increase in slip length that is in competition with an increase in water viscosity for nanometric pores. According to the authors' claim, it is a layered configuration of a water H-bond network in front of and behind the graphene nanopore that can lead to a decrease in the cross-pore water diffusion coefficient, ultimately increasing the water viscosity during the passage. The layering may thus reduce the probability of a water molecule to transmit from one water layer into another, thus resulting in an increased water viscosity.

Another study focussed on water transport across graphyne, a related structure to porous graphene, found an even stronger increase in the water viscosity that is inversely proportional to the sixth power of the graphyne pore characteristic length and attributed to attractive forces stemming from the Lennard-Jones potential that induces stronger H-bond orientation within the first few water layers directly over the membrane surface.[38] Even though the earlier report by Suk and Aluru[35] showed qualitative results of water viscosity in nanoconfining graphene pores as well, quantitative scaling differs between these two studies, which demands further investigation for the exact scaling and mechanism and also for the validity of the water viscosity alteration hypothesis itself.

The effect of pore functional group, which had been excluded in the previous study, was taken into account by Cohen-Tanugi and Grossman for pore sizes in the subnanometric regime possibly suited for desalination.[39,40] Hydrogenated and hydroxylated pores were compared. It was found that hydrogenated pores transport water significantly less than the hydroxylated pores for a comparable open pore area. They attributed this finding to a smaller chance of H-bonding between transporting water molecules and the hydrogenated pore edge such that the water molecules transfer across the pore in a more ordered manner. This picture can be interpreted as an entropic barrier imposing an (activation) energy penalty for the water passage across the subnanometric pores of graphene.

Water transport across graphene pores in the continuum pore-size regime was probed experimentally by Celebi, Buchheim *et al.* who prepared precise pore sizes from 50 nm to 1000 nm *via* FIB drilling on freestanding graphene.[7] They demonstrated that the unique transport properties of a double-layer CVD graphene membrane cannot be described accurately by the HP formalism, but instead, the transport physics follows the model proposed by Sampson. They were able to demonstrate that resistance to water permeation across the pore is accurately described by considering the entrance resistance alone, while pore passage resistance is negligible, a conclusion in agreement with the Sampson's formula. Mathematically, the pore-size scaling of the cross-2D-opening transport follows the third power of the pore diameter instead of the fourth power that the HP formalism demands.

Figure 3.2 shows measured, simulated, and analytical predictions of the per-pore-permeance of water across nanopores in porous 2D barriers. Sampson's theory can be seen to describe transport accurately for all pore sizes, while the H–P equation significantly deviates from experimental values for any pore larger than a few nanometres.

3.3.2 Ion Transport Across Nanoporous Graphene

The study of the transport of an ionic solute across nanoporous graphene membranes is relevant in various technologies such as desalination, biomedical applications, batteries, and fuel cells. Therefore, significant efforts have been made to understand and control ionic solute passage or rejection with nanoporous graphene membranes.

The first computational study in 2008 considered the transport of an ion solute across two distinct subnanometric graphene pores, one with a hydrogen-terminated edge and the other having an edge terminated with fluorine (F) and nitrogen (N).[41] Based on an electronegativity argument, the hydrogenated pore is expected to be slightly positively charged, while the F–N functionalization would cause negative charges to face the pore centre. Under an external electric field, the F–N functionalized pores show cation selective transport, and the hydrogenated pores are anion selective. Transport of various ions shows transmission rates consistent with the hydration radii of the ions considered (Li^+, Na^+, K^+, F^-, Cl^-, Br^-), indicating possible size selectivity. Electrostatic attraction determines whether the respective ion–pore combination enables entry and passage through a given pore, while the transport efficiency (or rate) depends on the size and the energy penalty associated with the hydration shell of each ion. Consequently, this study revealed the potential of a subnanometric graphene pore to act as an ion gate. A later study extended the analysis to larger pores up to 2 nm in diameter and thus simplified it by ruling out the size selectivity.[42] Still, the authors observed ion selective transport in these sub-2-nm pores if the pore edge is charged properly. Negative charges will facilitate cations while impeding the passage of anions, in agreement with the previous study. This behaviour is reminiscent of the Donnan exclusion theory in which counterions have a higher concentration within the membrane than in solution due to electrostatic interaction between the ion and pore charges. Interestingly, the ion selectivity trend is still maintained even though the Debye

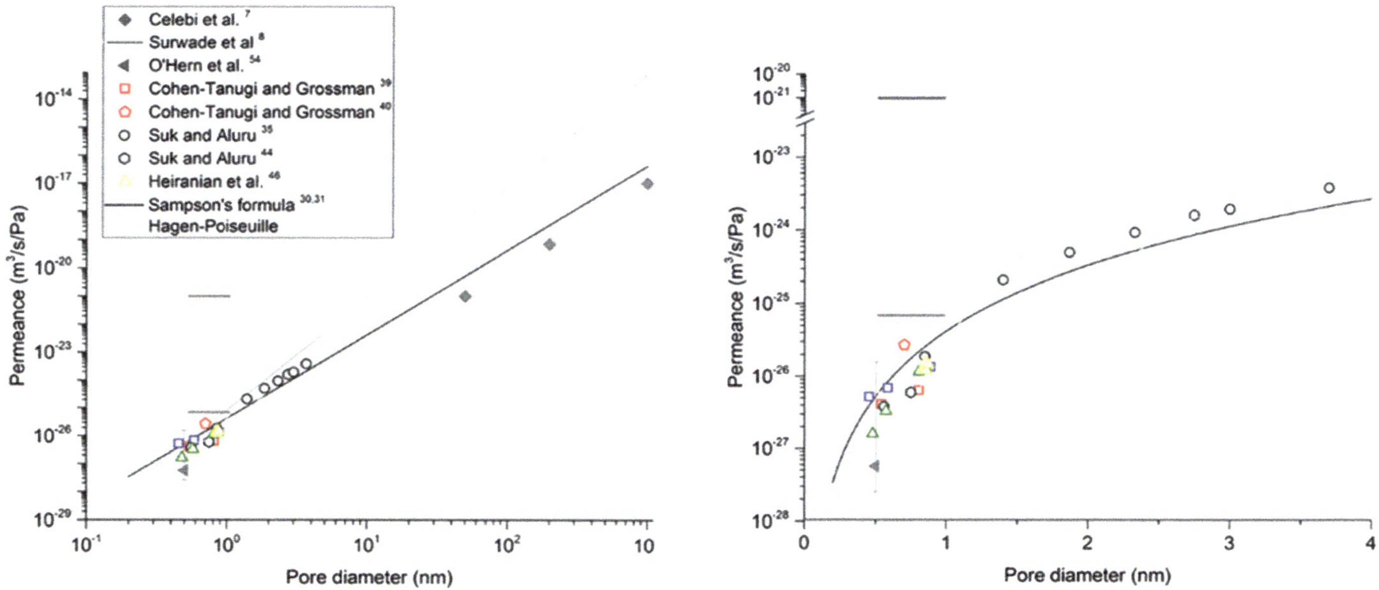

Figure 3.2 Per-pore-permeance as a function of pore diameter of various simulation and experimental data. Sampson's formula is accurate for all pore diameters, contrary to the Hagen–Poiseuille formalism. Symbols represent the corresponding study; colours represent chemical group at the pore edge. Black and red represent no and hydrogen functionalization, respectively. Grey represents unknown functionalization. Dark red represents molybdenum edge atoms and green represents mixtures of molybdenum or sulphur atoms. Green and dark yellow represent Mo only and Mo–S-mixed pore edge atoms in MoS_2. A water density of 1 g cm^{-3} is assumed unless the respective study provides more information.

screening length is much smaller than the pore diameter (~0.3 nm), raising a question that calls for additional investigations.

Analysing the ion transport across non-functionalized, uncharged pores with sub-5-nm diameters, Suk and Aluru looked into ion conductance and mobility through MD simulation.[43] They reported physical chemical aspects of aqueous ion solutions within these pores. In detail, they found that the number of water molecules within the first hydration shell of K^+ and Cl^- ions remains constant for pore sizes up to 1 nm. Smaller pores show a reduced coordination number that is attributed to steric exclusion and dehydration effects, however the coordination number reduces only slightly, indicating that a passing ion can compose its hydration shell from water molecules from both sides of the atomically thin graphene membrane. Simultaneously, the pore conductance decreases due to both causes: a lowered concentration of ions within the pore and a reduced ion mobility near the pore. The lowered concentration originates from a combination of steric hindrance and hydration energy penalty, while the ion mobility is reduced for ever smaller pores due to impeded ion diffusion. The impeded ion diffusion is attributed to the water layering configuration on both sides of the pore, as reported in a previous simulation.[44] These diameter-dependent concentration and mobility factors collaborate to change the nanopore's ion conductance from the value predicted on the basis of the bulk properties, if the pore diameter approaches a subnanometric dimension. For pores smaller than 9 nm in diameter, the continuum model description becomes erroneous such that their empirically derived diameter-dependent properties should be applied for an accurate description of graphene pore conductance.

Insight into selective ion passage across subnanometric graphene pores can also help to understand biological ion channels, as shown by He *et al.*'s MD simulation.[45] By mimicking biological ion channels by way of attaching carbonyl or carboxylate groups to the pore edge, it is possible to achieve preferential K^+ transport over Na^+ as is seen in biological ion channels under applied bias. Under applied voltage bias, K^+ is coordinated by one carbonyl group more than Na^+ and thus it is selectively transported, as the experienced passage barrier of K^+ is smaller than that of Na^+. Furthermore, Na^+ is observed to bind more strongly to carboxyl groups than K^+, leading to preferential Na^+ passage over K^+ in the smallest pore at low voltage, since the bound Na^+ blocks K^+. Na^+ is transported by a knock-on mechanism and due to the stronger Na^+ interaction with the pore-edge functional groups, it is more likely to be replaced by a new Na^+ rather than a K^+. Therefore, Na^+ selectivity can result. At high voltage, however, Na^+ cannot block the pore anymore, since the carboxyl group reacts to the larger electric field by swinging out of the graphene plane. Then, K^+ ions are not blocked from passage anymore, and since Na^+ is attracted strongly to the pore edge, its permeation rate decreases.

Graphene's potential to selectively transport ions also raises the question about complete ion rejection that would imply a membrane feasible for desalination applications. Cohen-Tanugi and Grossman delved into this question by means of MD simulation for various subnanometric pores functionalized with OH^- or H^+.[39,40] For effective pore diameters above 0.55 nm, the pores lose their rejection capability, but below that threshold, selectivity up to 100% can be achieved. At a given pore size, hydroxylated pores show lower rejection of salt ions, possibly attributable to the H-bond between hydroxyl groups and hydration-shell water; the free energy barrier could be lowered if water molecules from an ion hydration shell could be replaced by a hydroxyl functional group at the pore edge during the passage, a mechanism the hydrogenated pore is missing. Interestingly, the subnanometric graphene pores lose the ion rejection capability as pressure increases. The authors hypothesized that the larger the effective volume of an ion hydration shell, the more sensitive the graphene pore becomes to a pressure increase, however, facilitated dehydration of salt ions during passage at higher pressures may also explain this observation. Furthermore, the simulation conditions render water in the compressible regime, which may have an effect on the salt hydration layer. Still, linear extrapolation of salt rejection of a 0.8-nm-wide pore to practical pressures (*e.g.*, 5 MPa) results in nearly 100% salt rejection, which raises the question of how salt rejection and pore size are related at lower pressures and the possibility of desalination with pores larger than that stated in this study.

Apart from graphene, other 2D materials show similar promise for high permeation due to their thinness. An MD study of water desalination across nanopores in monolayer molybdenum disulphide (MoS_2) in fact revealed 70% higher permeance than that across graphene pores.[46] The unique structure of MoS_2 allowed for comparison of three different pores with Mo-only, S-only, or a mixture of those atoms to be located at the pore interior. A pore with Mo edge atoms shows the fastest transport, while pure S atoms are the slowest transporters of water. A difference between these pores is that Mo pores have an hour-glass cross section due to the atomic structure, while S pores do not have this geometry. The authors attribute the fast flow to this hour-glass geometry and furthermore to hydrophilicity of the Mo sites. However, the application of the concept of hydrophilicity and hydrophobicity at the atomic level is questionable, as the hydrophilicity of materials at the macroscale can change significantly under nanoconfinement or structuring. Although the authors agree with Cohen-Tanugi and Grossman in that faster permeating pores reject less ions, their simulation shows around 95% rejection for both MoS_2 and non-functionalized graphene, while Cohen-Tanugi reported merely ~40% rejection for similar pore sizes and pressures.[39,40] Such a remarkable difference demands further investigation of the true rejection capability of atomically thin membranes for a given pore size, pressure, and chemical functionalization. The authors further extended the analysis to other transition metal dichalcogenides to observe that mainly the metal component dominates the permeation and salt rejection capabilities.

The first experimental characterization of ionic transport across nanoporous graphene was published in 2010 by Garaj *et al.*[47] They showed the nanopore conductance of CVD graphene to be almost linearly increasing with diameter. From their conductance measurements with various pore diameters, it was possible to extract the effective insulating membrane thickness to be 0.6 nm, a value that was confirmed by DNA translocation measurements with a ~5-nm-wide pore and matches well with the theoretical work of Suk and Aluru.[43] Using the DNA translocation experiments, they could show subnanometre resolution of graphene for DNA discrimination, rendering atomically thin porous graphene a promising candidate for DNA sequencing applications.

Extending the analysis of the ion conductance to pores up to 2 nm, Jain *et al.* characterized a current-voltage behaviour across intrinsic defects of CVD graphene, in resemblance to biological ion channels.[48] A Nernst–Planck model incorporating electrostatic and steric interactions of a graphene pore of variable diameter and charge with a single ion could allow the transport properties of the pores to be extracted. Linearity between current and voltage is attributed to uncharged pores with diameters above the hydrated diameters for the ions such that a charge-neutral pore with pore size larger than the ion hydrated radius cannot impose steric hindrance on the passing ion. Voltage-activated behaviour could be attributed to uncharged nanopores that sterically hinder the passage of the hydrated ion. The nonlinear current increase above a certain threshold voltage may originate from increased ion dehydration due to the higher electric-field driving force. At small electric fields, the dehydration barrier strongly impedes ion passage leading to small currents. The presence of charge at the pore mouth may alter the transport in two possible ways: a charge that is positioned symmetrically could result in the current-saturation behaviour by imposing electrostatic repulsion on passing ions, while an asymmetric placement of the charge at the pore mouth yet out of the pore plane could produce a rectified current–voltage characteristic. Here, the asymmetric placement may couple the perceived near-pore electric potential toward the direction in which an ion permeates, causing an asymmetric current–voltage behaviour. In some cases, rapid current fluctuations are observed for samples that otherwise show the voltage-activated behaviour. This current fluctuation is attributed to a protonation–deprotonation transition because of time-scale similarity and power spectra analogous to those obtained for the protonation–deprotonation transition. What is not clear, though, is the otherwise voltage-activated behaviour for these devices that was previously attributed to steric hindrance in the absence of charge, while protonation induces pore charge alteration. This seemingly conflicting behaviour calls for further examination. Still, the remarkable similarity to transport in biological channels lends graphene nanopores eligibility as a model platform to mimic biological ion channels.

The smallest nanopore thus far probed for ionic transport is ~0.3 nm in diameter perforated into single layer MoS_2 membranes.[49] Opening pores in

MoS_2 by electrochemical means allowed for controlled fabrication of single pores in suspended MoS_2. For a 0.6-nm-wide nanopore, current–voltage characteristics show negligible current below a certain threshold voltage upon which the current non-linearly increases. The voltage range in which current is suppressed depends on the pore size and the cation valence that passes through the pore. The nanoelectronics concept of a Coulomb blockade is applied to ionic transport to explain the observed phenomena. This model dictates that an individual ion can block a nanopore, stopping other ions from passing through it, yielding negligible current at a small bias. Increased voltage bias can release this blockade through a mechanism whereby augmented electrostatic interaction between cations in the vicinity of the nanopore eventually removes the cation from the pore. Negative charges at a nanopore could cause cations to bind to it, resulting in current blockade. By changing the pore size, the authors argued that a dehydration energy penalty of ions dominates the current blockade for sub-0.6-nm-wide MoS_2 nanopores while justifying an ohmic current–voltage behaviour for super-1-nm-wide nanopores that are too large for a single ion to block them. Their observed current-voltage characteristics also resemble biological ion channels of dimensions similar to their nanopores.

These researchers furthermore employed concentration-gradient-driven selective ion diffusion across an individual nanopore on monolayer MoS_2 to demonstrate a power generator.[50] With pores ranging from 2 to 25 nm, selective ionic passage could be achieved, attributed to negative surface charge evidenced by conductance saturation at low concentration and increased conductance at higher pH. The surface charge is capable of screening anions due to the significant Debye length (~10 nm) resulting in a net positive current following the concentration gradient. The smaller the pore, the higher the ion selectivity of the nanopore, resulting in a larger voltage generated at the expense of smaller current. Similarly, the lower selectivity of a larger pore induces lower voltage yet at larger current, a trade-off that hints at an optimal nanopore size to maximize power generation. If a membrane bears nanopores with an average pore diameter of 10 nm at 30% areal porosity, it may potentially generate as great a power density as 1 $MW\,m^{-2}$ thanks to efficient transport across the atomically thin membrane. However, a sub-additive ion–current increase may be expected as well for 2D porous membranes, as has previously been observed for solid-state nanopores.[51]

While the role of surface charge on MoS_2 that screens ion passage is based on the Debye layer thickness, Rollings *et al.* probed selective ionic transport across graphene nanopores where the Debye layer is much smaller than the pore size.[52] For up to 50 nm-in-diameter pores, K^+-to-Cl^- selectivity neared almost 100 and relies on the solution/electrolyte pH, suggesting a mechanism of protonation–deprotonation of chemical moieties at the pore edge. A pH scan revealed that the graphene nanopore edge is negatively charged at neutral conditions. However, since the Debye screening length (1 nm) is much smaller than the pore diameter and charge screening alone from the pore edge cannot explain the observed selectivity, another mechanism for selectivity needs to be conceived. Ion selectivity of pores much larger than the Debye screening length is rationalized by the negative surface charge of graphene that can attract a mobile cloud of screening cations. These mobile cations can diffuse along the graphene surface and cross around a pore edge, causing net ionic current to be cationic. This hypothesis is supported by conductance measurements at pH 8 and 2 that show higher transmembrane ion conductance at pH 8, indicating that more cations screen the negative surface charge of graphene than in the pH 2 case. Numerical solutions of Poisson–Nernst–Planck (PNP) equations for a surface charge density of null and $-0.6\ \mathrm{C\,m^{-2}}$ confirm that results of the strong negative charge case agree with the measurements well. In comparison to Feng *et al.*, Rollings *et al.* reported that an approximately 10-fold higher surface charge may result in great selectivity even at a pore diameter of 50 nm.[50] The origin of the surface charge is not fully understood but may be related to their sample preparation method involving voltage pulsing, given that other researchers have reported much weaker selectivity at smaller pores.[9]

Complete ion rejection across subnanometric pores of monolayer graphene has been reported by Surwade *et al.*, implying the potential of graphene as a water desalination membrane.[8] Subnanometric pores could be created by O_2 plasma treatment, as evidenced by aberration-corrected scanning transmission electron microscopy. By contacting one side of the graphene membrane to deionized water at 40 °C, they measured extremely high mass flux across the O_2-plasma-treated graphene membrane, which is surprising and calls for further investigation of phase change under nanoconfinement. Subsequently, ionic current measurement of this membrane (0.5–1-s-long O_2 plasma treatment) obtained none to very low ion conductance, indicating ion rejection and the potential as an effective desalination membrane. Repeating the initial permeation experiments with an ionic solution shows much slower water permeation and almost complete ion rejection for <1 s plasma treatment time. Osmotic water flow experiments further revealed the semipermeable nature of the membranes and their desalination capabilities. The osmotically driven water transport rate occurs exclusively in the liquid phase and matches well with theoretical predictions.

Any real membrane for liquid-phase separation applications is likely to be grown by CVD and of macroscopic size. Membranes prepared in this way often accompany intrinsic defects such as pinholes from graphene growth or ruptures from membrane fabrication. As was similarly probed for gas transport, O'Hern *et al.* investigated transport across intrinsic defects of graphene over a macroscopic area of $O(10)$ mm.[27,53] They found an inhibited transport of TRMD (*ca.* 12 nm in size), while those of KCl, allura red, or TMAC were not significantly affected, when the permeance values were corrected for uncovered bare PCTE areas. This reveals the presence of intrinsic defects of 1–15 nm in size, as is confirmed by electron micrographs. Control over subnanometric pore sizes in a potentially scalable process was achieved in a later work using first energetic ion bombardment of CVD graphene with subsequent wet chemical etching to grow defects into permeable pores.[9] For a certain etching time, it was possible to demonstrate enhanced KCl transport compared with allura red, supporting a size exclusion argument and indicating the generation of pores larger than KCl (0.7 nm) but smaller than allura red (1 nm).

Since macroscopically grown CVD graphene often contains both intrinsic defects and ruptures originating in membrane handling, a two-stage strategy can be applied to seal membrane defects.[54] In the first stage, atomic layer deposition (ALD) of *ca.* 3.5 m thick hafnia (HfO_2) was applied targeted to close defects below 15 nm in size, after which interfacial polymerization (IP) of nylon-(6,6) was utilized for sealing larger defects that originate from ruptured graphene on a 200-nm-pored PCTE support. Subsequent etching by potassium permanganate can create pores with an average value of 0.16 nm, as analysed by HRTEM, with a small fraction of pores larger than the water van der Waals diameter and rarely salt permeable pores (>0.7 nm). Osmotic pressure experiments show water transport close to theoretical prediction and of a similar order of magnitude as polymeric membranes typically applied in reverse osmosis. Solute rejection was studied with NaCl, $MgSO_4$, allura red, and dextran showing negative rejection of NaCl (0.7 nm), and rejection of 70% $MgSO_4$ (0.8 nm), 90% allura red (1 nm), and 83% dextran (3.7 nm). Negative NaCl rejection is equivalent to more NaCl transport through their graphene/PCTE composite in comparison to the bare composite, according to the used rejection definition. This observation along with the lower dextran rejection despite the larger size requires further study and might be related to the KCl permeability of nylon.[55] Further, the role of pores too small for IP sealing and too large for ALD sealing remains unclear.

To target applications properly, achievable salt rejection as a function of the pore diameter needs to be understood. Figure 3.3 shows various experimental and theoretical results for various nanopores. In general, pores below 0.5 nm are found to reject salts such as NaCl, however for pores slightly larger than this size, the rejection decreases but also becomes inconsistent within a pore size. While some research predicts over 98% rejection of *ca.* 0.85 nm pores, others simulate down to a mere 30% rejection at a given pore size. Differences in chemical functionalization of the pores, but also in simulation details, may cause this deviation. Furthermore, experimental demonstration of NaCl rejection using CVD graphene remains to be achieved as the only experimentally available result shows the peculiar case of negative rejection that may be related to the experimental procedure and data evaluation of that work.

In summary, liquid and ion transport across graphene has shown diverse behaviour in both theory and experiment that strongly varies with pore size, charge, and transporting solutes. The demonstration of selectively passing cations over anions paves the way for efficient reverse electrodialysis, and the demonstration of ion rejection is particularly important for the desalination

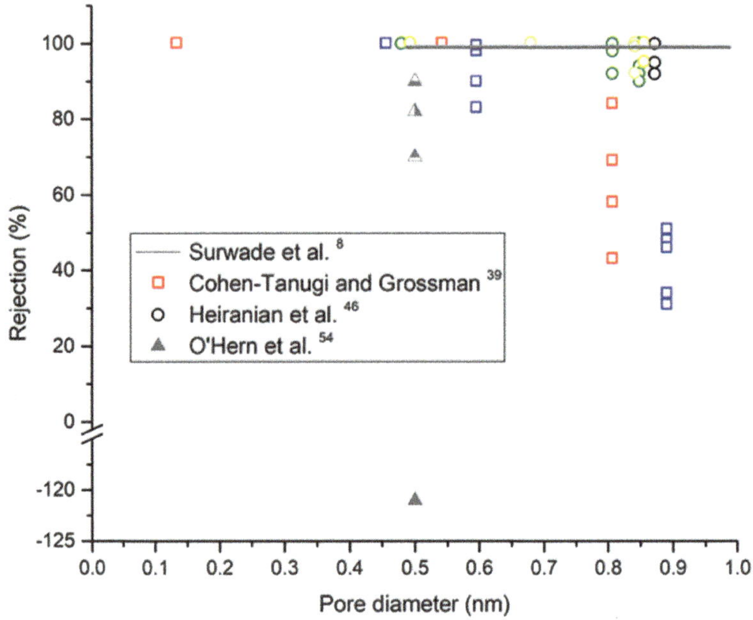

Figure 3.3 NaCl rejection, unless specified differently, as a function of pore diameter. Inconsistent ion rejection at a given pore size needs further investigation. Symbols represent the corresponding study; colours represent chemical group at the pore edge. Red, blue, black, and grey represent hydrogen, hydroxyl, no, and unknown functionalization of graphene, respectively. Green, yellow and dark yellow represent Mo, S, or mixed pore edge atoms in MoS_2. Partially filled triangles represent $MgSO_4$, allura red, and dextran molecules.

applications. Substantial efforts are anticipated and required in obtaining selective molecular flow across graphene membranes at square centimetre or larger scales, which calls for a breakthrough in the manufacturing process. On the other hand, applications that require only a micrometre-scale membrane dimension, such as DNA sequencing or chem-bio sensing, might have more readiness for market entrance.

3.4 Proton and Atomic Species Transport Across Atomically Thin Membranes

Miao *et al.* used first principles DFT simulations to study atomic hydrogen and proton transport through pristine graphene.[56] Due to electron orbital overlap, repulsion forces are experienced by both hydrogen and protons. However, the energy barrier for passage of physisorbed hydrogen is 2.46 eV compared with physiosorbed protons experiencing only 1.41 eV. During passage, the graphene hexagons expand and contract again, reflecting the strong interaction, similar to gas molecules passing through nanopores.

This study demonstrates the principal possibility of hydrogen and proton passage through pristine graphene, however, the predicted energy barriers would be impractically high. The presence of defects such as double vacancies, on the other hand, significantly reduces the energy barrier for passage.

Experimentally, Hu *et al.* investigated the proton conductivity of mechanically exfoliated monocrystalline graphene, MoS_2, hBN, and multilayers of these materials with an applied electrical potential as the driving force.[57] By coating different 2D crystals with Nafion™ and measuring the current–voltage characteristics in a humid atmosphere, they found unexpectedly high transport through graphene and hBN monolayers, while single layer MoS_2 and bilayer graphene showed no transport. The variation in proton conductivity is rationalized by a variation in electron cloud density among the various materials as well as the difference between hBN and graphene in stacking behaviour, however, the fundamentals of the passage remain to be elucidated. Measuring proton conductivity at various temperatures allowed the passage energy barrier to be extracted, which was determined to be 0.78 eV and 0.3 eV for graphene and hBN monolayers, respectively. Platinum sputter coating of the 2D crystal further decreased the passage barrier, leading to immense proton conductivity for hBN. Further experiments of HCl conductivity measurements and mass spectroscopy verified the previously observed transport rates. The lower passage barrier compared with theoretical prediction requires further research.

Measuring aqueous proton transport across graphene in the absence of an external electric field as the driving force was reported by Achtyl *et al.*[58] Using alternating streams of acidic or basic solutions of the same strength flowing over a fused silica surface allowed them to probe the protonation and deprotonation of silanol at the silica surface by the second harmonic generation technique. The measured second harmonic time trace is indistinguishable for bare fused silica compared with fused silica covered by monolayer graphene from CVD, indicating unimpeded proton transport across the graphene layer. The same holds true for up to 8 layers of tested graphene. SEM analysis shows macroscopic defects of 500 nm and larger in diameter widely spread such that it is difficult to probe areas close to these defects due to the 30 μm wide laser spot. However, STEM analysis with atomic resolution reveals atomic scale defects to be always statistically present within the laser spot. DFT simulations and ReaxFF reactive force field MD simulations were performed to predict the aqueous proton transport mechanism across graphene. They obtained a 3.8 eV energy barrier for protons to pass through pristine graphene, which makes it an unlikely explanation of their results. However, for quad-vacancy atomic scale defects with various pore functionalization, they found that hydroxylated pores can allow proton passage with a moderate energy barrier of 0.6–0.7 eV as these can form hydrogen bonds with the adjacent water layers such that protons can shuttle across the defects in a Grotthuss mechanism at room temperature. Due to uncertainty of the atomic defect density estimation, the authors state that other means of transport may yet be possible. Especially, the reason for indistinguishable traces for even eight layers of graphene remains elusive.

In the same year, a second study measured proton transport across CVD graphene by covering glass capillaries with single layer graphene and measuring current–voltage characteristics in the presence of a HCl concentration gradient.[59] Selective passage of protons over Cl^- anions is observable by a measurable net current with no voltage applied. The reversal potential, the external potential required to stop the net current, allows the proton selectivity of the membrane to be extracted. While as-grown graphene shows only mild selectivity, depositing Al_2O_3 by ALD causes the total current across the graphene to decrease significantly, while at the same time, the reversal potential, and thus the selectivity, increases significantly. A reduction in total current and an increase in reversal potential after ALD coating is interpreted by the authors as proton transport through defects since the total current should be dependent on the defect size, which decreases with ALD coating, and increased selectivity could be caused by the reduction in defect size, such that they become more selective toward protons compared to Cl^-.

A theoretical analysis to resolve the discrepancy of the graphene tunnelling barrier between previous MD simulation and experiments was executed by Poltavsky *et al.*[60] They employed *ab initio* MD Feynman–Kac path-integral simulations that treat the atomic nuclei quantum-mechanically instead of classically. For comparison, they also carried out their simulation by treating the nuclei classically and they obtained similar transport barriers for the proton as earlier studies. The quantum mechanical treatment of the nuclei, however, predicts a significantly lower energy barrier for thermal protons to pass through graphene (0.6 eV) and this matched with that determined experimentally (0.8 eV) much better than the non-quantum-mechanical treatment of the nuclei. Thus, this theoretical study concludes favourably for the proton transport mechanism across pristine graphene.

Seel *et al.* investigated proton and atomic hydrogen transport across pristine 2D materials, such as graphene, hBN, MoS_2, and others by means of DFT to shed light on the question of proton transport through atomically thin 2D materials.[61] From their simulations, they found that system relaxation, that is, non-rigid 2D material atom positions, is a significant factor determining the penetration barrier height. Protons passing through graphene experience a 1.38 eV energy barrier, while hBN only poses a 0.11 eV energy barrier. Apart from slightly larger atomic bond lengths in hBN compared with graphene, the polarization of hBN is found to facilitate proton transport due to ionic bonding opportunities for protons arising during passage. Contrary to this result, MoS_2 is found to trap atomic hydrogen and protons in between the S-layers in a 1.56 eV deep energy well. They investigated the effect of Pt for passage without observing a different activation energy for passage. The presence of a quad-vacancy terminated by oxygen atoms reduces the passage barrier further to 1.1 eV, suggesting that experimentally observed proton transport across graphene stems from defects, while hBN allows thermal proton passage.

Zhang *et al.* carried out ab-initio DFT calculations to better understand the differences in permeance of hydrogen isotopes, as previously demonstrated by Lozada-Hidalgo *et al.*[62,63] They compared graphene, hBN, alpha-boron, as well as graphene with Stone-Wales (55–77) defects. Pristine graphene is predicted to have a 1.5 eV passage barrier, while hBN only imposes a 1 eV barrier, both in quantitative opposition to experimentally measured values. However, the ratio of hydrogen isotope separation, *e.g.* proton/deuterium, based on differences in passage barrier and Arrhenius rate constants amounts to 12, close to the experimentally obtained values. The same holds true for proton/tritium separation with a predicted selectivity of 37, close to the experimentally measured value of 30. They further probed the passage barrier of Stone–Wales defects, where specifically proton transport across heptagons is found to experience a lower energy barrier of 0.55 eV and an increased proton/deuterium selectivity of 25, which is attributed to stronger interaction of protons with C–C bond connecting pentagons of the Stone-Wales defect. Furthermore, alpha-boron as a model system was investigated to yield a proton passage barrier of only 0.2 eV. The differences in passage barriers for 2D materials is attributed to the electron-density surfaces that reveal graphene, hBN, and alpha-boron to have different effective pore sizes experienced by a passing proton, resulting in the difference in the passage barriers.

Another study extended the analysis of proton permeation across two-dimensional materials from graphene and hBN to other materials such as phosphorene or silicene.[64] In their DFT simulations, they furthermore distinguished the proton permeation barrier depending on the environment around the membrane. Under vacuum, protons pass through graphene and hBN with 1.2 eV and 0.6 eV energy barriers for adsorption with subsequent passage. Changing the environment to aqueous solution is found to increase the proton penetration barriers to 3.3 eV and 3.0 eV for graphene and hBN, respectively, such that the authors concluded proton passage to be unlikely at room-temperature in an aqueous environment.

More efforts to understand and resolve the mismatch between theoretical and experimental proton permeation barriers across graphene and hBN were made by means of DFT simulation.[65] Structural optimization of the proton-2D-crytal arrangement was found to counterintuitively increase the permeation barrier. This could be explained by the previously unstable configuration of the proton in vacuum over the 2D crystal. System relaxation leads to proton adsorption to the respective crystal lattice such that the system is in a more stable state. Other effects of mechanical strain or curvature reduce the barrier by 0.1 eV or less and thus are unable to explain the discrepancy. Further simulations of a proton in aqueous phase also increased the permeation barrier to up to 5 eV for graphene, in qualitative agreement with the previous result[65] of a higher barrier to solvated protons in comparison to vacuum permeation.

In a study performed by Tsetseris *et al.*, DTF was applied to reveal details of the permeation process of atomic hydrogen, boron, nitrogen, and oxygen.[66] With passage energy barriers of 4.2 eV, 5.5 eV, and 3.2 eV for hydrogen, oxygen, and nitrogen, respectively, it was confirmed that pristine graphene can be considered impermeable also for these atomic species. Interestingly, the passage barrier for atomic boron is only 1.3 eV. In general, three passage ways for permeation of atomic species can be considered: direct passage through the centre of the hexagonal rings without C–C bond breaking, adsorption and passage involving breaking and reformation of C–C bonds, and more complex passages. Using minimum energy pathways, the authors determined all of the studied atomic species to predominantly undergo the second pathway involving breaking of C–C bonds by forming bonds with the passing atoms. Out of these, boron has a surprisingly low energy barrier, which may be overcome frequently at moderately high temperatures of around 200 °C.

Experimentally, selective hydrogen isotope transport was demonstrated by means of a Nafion™ coating of graphene or hBN monolayer in a 100% humidity atmosphere of either hydrogen–argon or deuterium–argon mixtures and then by performing current–voltage measurements and mass spectroscopy measurements.[63] The authors found a proton/deuterium selectivity of 10, which can be interpreted as a 60 meV difference in passage energy barrier. Interestingly, the selectivity is the same for graphene and for hBN, even though the total barriers differ. This result implies intrinsic differences between protons and deuterons to cause the observed selectivity. Indeed, the differences in zero-point energy of protons bound to oxygen of the SO_3^- group of Nafion™ from those of deuterons matches the derived energy difference from the selectivity. This finding suggests the difference in zero-point energy is responsible for the experimentally observed selectivity. From the differences in zero-point energy, a proton/tritium separation factor of 30 is expected.

In a subsequent work, Lozada-Hidalgo *et al.* demonstrated the electrochemical pumping approach for CVD graphene on a one-inch scale with 95% macroscopic graphene coverage and a proton/deuteron separation factor of 8.[67] Since graphene works as an electrode of the electrochemical pump, these macroscopic pinholes are not expected to reduce the separation factor significantly, validating the utility of the technology. The energy requirements are less than the currently best available technological option showing the highest current selectivity and may be reduced further by using hBN and/or optimized operation conditions.

Despite these advances, the fundamental question about proton transport across pristine graphene persists, as can also be seen from Figure 3.4 showing the available results of energy barriers for protons to pass through graphene as a function of the number of missing carbon atoms from the lattice. An apparent mismatch between simulation and experiment can be observed. Particularly, the presence of atomic defects with a few atoms removed from the lattice are experimentally hard to rule out unambiguously.

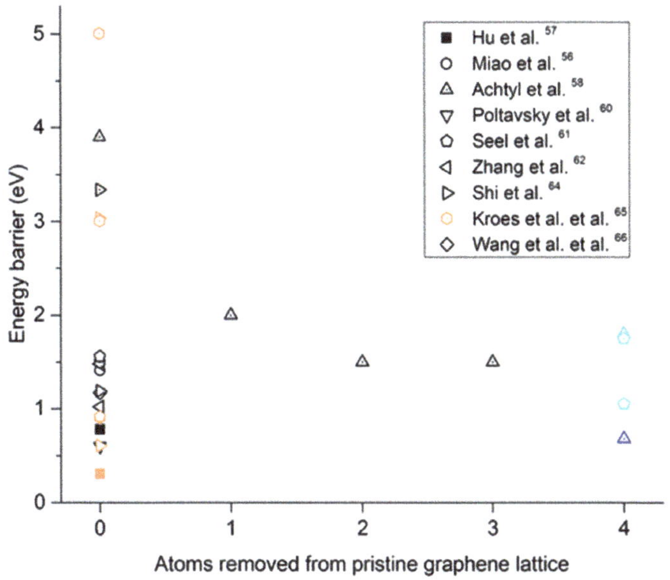

Figure 3.4 Energy barriers for proton passage across graphene and hBN for various atoms removed from the lattice. The strong variation between available data for non-porous graphene transport barrier requires further study. Symbol shapes represent different studies. Filled and hollow symbols represent experimental and theoretical results, respectively, with black and brown representing graphene and hBN, respectively. Dotted symbols represent an aqueous environment. Blue and cyan represent graphene with hydroxyl and oxygen termination. Kroes *et al.* provided a range of penetration barriers in between the limits shown in the graph.[65]

3.5 Conclusion

Chemical species can transport across porous graphene or other two-dimensional membranes through various mechanisms depending on the phase and charge of each transporting species. Simulations and experiments agree that pristine graphene is generally impermeable to gases, while nanometre-scale pores punctured on graphene can provide transport pathways depending on the size relation, physicochemical interaction, and orientation of transporting molecules with pores, on the geometry and chemical functionalization state of the pore, and on the presence of non-permeating species. While the molecule-to-pore size ratio can primarily determine passage, a precise amount or rate of permeation of a gas species may be reliant on surface adsorption and diffusion or the chemical affinities of the species around the pore, factors often regarded as secondary effects to porous membranes. According to simulations, electron orbitals and the resulting electron probability density can effectively govern the interaction dimension between pore and molecule. From this finding, it is deduced that electron orbital overlap is strongly correlated with the energy barrier that molecules face in an attempt of pore passage. For a certain pore size near 1 nm (slightly beyond the molecular sieving regime), adsorption to and surface diffusion on the membrane surface are likely to dominate the gas transport mechanism. This prediction may have significant implications for membranes in practice as a typical separation prediction based on molecular mass or kinetic diameter may not work for certain gas mixtures, possibly resulting in separation factors that belie those of molecular sieves, as exemplified by favourable permeation of unwanted species. Pores in the size regime tightly commensurate with the transporting molecular dimension are likely to separate a gas mixture *via* entropic gates that facilitate certain specifically oriented molecules. This approach may prove useful in the separation of gas species that are close in kinetic diameter but disparate in adsorption orientation. An idealization taken in many simulations is membrane rigidity, whose insignificant influence on separation has been put into question. In this light, experimental validation of predicted phenomena can be critical to clarify our molecular-level understanding of gas transport and separation across porous 2D materials. Similarly, no experiment to date has been able to reveal the disparity in separation correlated to functional groups at the pore edge and their charge state. For molecules of very similar kinetic diameters, the clear disparity in molecular interaction with pore functional moieties promises to allow meaningful separation of an almost inseparable mixture with conventional membranes, such as N_2 from O_2.

Experimentally, hermeticity and molecular-sieving-like large separation factors ($>10^4$ for a certain gas mixture) of mechanically exfoliated graphene have been established. Transport across exfoliated graphene with few pores is found to be time-variant, originating possibly from dynamic switching of the bond state of edge atoms of subnanometric pores. Flow physics of large-area CVD-grown graphene does not follow continuum-mechanics-based channel-flow models such as Hagen–Poiseuille flow but instead complies with Sampson's formula, a solution of a low-Reynolds-number flow across a 2D disc, where resistance to flow comes exclusively from a pore entry event. For the same reason, no Knudsen minimum in the permeance is observed at Knudsen numbers near unity. Regarding gas separation, a critical role of defects in graphene has been verified. Molecular sieving across defective graphene with subnanometric pores has been characterized in a single-gas permeance measurement followed by estimation of permselectivity. However, real applications demand sieving of gas mixtures, calling for methods to generate subnanometric defects in a facile, controllable, and scalable manner, while maintaining the ultimate permeance promise of graphene.

In order to apply the graphene- or 2D material based membranes to gas separation applications, there are a number of physical effects and technological aspects to account for. Demonstration of gas mixture sieving across CVD graphene membranes perforated with scalable processes needs to be achieved to move CVD graphene membranes from the laboratory to industry for applications such as air separation, CO_2 sequestration or other technologically relevant gas separations. Physical effects of charge and chemical functionalization of a pore remain to be investigated, and the predicted blockage of non-permeating species in ternary mixtures defines both a scientifically interesting question – for the exploration of new means of separation – and a practically important task if considering a fact that most gas separation applications deal with complex mixtures beyond binary ones. Regarding system design, both ultrahigh selectivity and permeance of membranes may at some point add negligible performance improvement to an entire gas separation system owing to such external effects as concentration polarization, retentate recycling limitations, and so forth. With 2D membranes offering ultimate permeation, it may be possible to reach the technologically meaningful limit of permeance. Given a proper way of producing large membranes with subnanometric pores, the same may hold true for selectivity. Then, these membranes can be considered as the ultimate membranes, since further permeance or selectivity improvement of the membrane may not necessarily lead to deterministic system improvements; an ultimately permeable separation membrane is not the limiting factor anymore.

Regarding liquid transport, Sampson's formula governs the transport across a 2D aperture as long as continuum fluid can be assumed around the 2D aperture, although frequently, the Hagen–Poiseuille formalism is misused for subnanometric pores because both theories lead to comparable predictions for channels with pores of aspect ratio close to unity. Variation of permeance for different 2D materials such as MoS_2 has been theoretically predicted, though experimental confirmation is absent. Pore-sizes below the limit of continuum assumption have been predicted to exemplify sub-continuum variation in fluid properties such as density, viscosity or diffusivities for which experimental proof is still missing. Vapour transport studies across nanopores have so far been limited, however interesting questions about fluid properties during phase change and vapour interaction with graphene or other 2D materials can be thought of due to similar experimental observations in neighbouring fields.[68] The understanding of flow enhancement for CNT and nanofluidic channels may be broadened by considering transport across 2D nanopores that are at the limit of thinness and could present an idealized system of the former fields. Even though graphene has been proven as mechanically sturdy under applied pressure as simulation and previous mechanical characterization predicted, the question of how the mechanical properties will alter if graphene is perforated remains.[69,70] For use in practical applications, it will be necessary to probe the limits of mechanical strengths for various pore sizes, density and other membrane parameters, as has been predicted theoretically.[71]

Ionic transport has been investigated rather thoroughly and theoretically where the conductance of 2D nanopores decreases strongly at the continuum-limit, due to non-continuum effects such as variation in ion mobility, concentration, or dehydration barriers for passage. Nanopores with different functional groups exhibit a strong ion selective behaviour and a possible inter-co-ion selectivity depending on interaction strengths with charged groups at the pore edge. For sufficiently small pores, desalination by reverse osmosis is predicted and experimentally shown using forward osmosis. Similar to pure liquid transport, distinct effects of the choice of 2D materials is predicted, although experimental proof is lacking. Experimentally shown, however, have been various ionic conductance phenomena of subnanometric pores such as Coulomb blockade or conductance similar to biological channels. At large pore sizes, linear scaling of conductance with pore diameter is most accurate due to the 2D geometry of the pore. This scaling is in line with the liquid transport where the transport rate shows a linear dependency on the pore size (*e.g.*, Sampson's formula), in contrast to three-dimensional descriptions. Besides, molecular sieving using centimetre-sized graphene and nanofiltration of charged species are established.

The effect of surface charge has been proposed as a significant means for selective ion transport even in pores much larger than a Debye screening length, raising the question of how to tailor the surface- and pore charge in order to engineer ionic transport for various pore sizes. Variation in surface charge by voltage gating may offer a pathway to achieve ion selectivity for pores larger than the hydrated diameters of the solutes to be rejected. For desalination applications, salt rejection by CVD-grown graphene perforated with scalable processes remains to be demonstrated in order to position graphene as a potential candidate for future desalination membranes. Proton transport as a special case of ionic transport due to differences in atomic configuration in solution has been experimentally shown and theoretically analysed, however the exact mechanism of transport remains disputed with some studies suggesting proton transport through pristine parts of graphene or hBN, while others suggest defects as the cause of the proton transport pathway. Here, more work is required to resolve the current issues and to enhance our understanding.

As an overarching goal, 2D membranes should be manufactured at length scales relevant to the respective application, which often involves square metres or even larger than that. To this end, synthesis and fabrication methods to yield nearly defect-free membranes need be established. One of the most important operational challenges in membrane separation processes is clogging and fouling of the membrane surface such that separation performance of the membrane deteriorates over time. 2D membranes might offer a unique advantage of reducing clogging and fouling exclusively to their surface compared with channel clogging and fouling in conventional membranes. Surface cleaning should be much less challenging, and thus 2D membranes may prove themselves to be practically beneficial for long lifetimes under chemically harsh conditions.

Acknowledgements

We acknowledge the financial support from Commissions for Technology and Innovation (CTI), Federal Department of Economic Affairs, Education and Research (EAER), Switzerland (KTI-Nr. 18463.1 PFEN-MN). A part of this work was financially supported by the Korea Institute of Energy Technology Evaluation and Planning (KETEP) and the Ministry of Trade, Industry, & Energy (MOTIE) of the Republic of Korea (No. 20168510011420) for which H.G.P. is grateful.

References

1. H. B. Park, J. Kamcev, L. M. Robeson, M. Elimelech and B. D. Freeman, *Science*, 2017, **356**(6343), eaab0530.
2. R. W. Baker, *Membrane Technology,* John Wiley & Sons, Inc., 2012.
3. M. Elimelech and W. A. Phillip, *Science*, 2011, **333**(6043), 712.
4. B. Freeman and Y. Yampolskii, *Membrane Gas Separation,* John Wiley & Sons, Inc., 2011.
5. J. S. Bunch, S. S. Verbridge, J. S. Alden, A. M. van der Zande, J. M. Parpia, H. G. Craighead and P. L. McEuen, *Nano Lett.*, 2008, **8**(8), 2458.
6. M. D. Fischbein and M. Drndić, *Appl. Phys. Lett.*, 2008, 93(11), 113107.
7. K. Celebi, J. Buchheim, R. M. Wyss, A. Droudian, P. Gasser, I. Shorubalko, J. I. Kye, C. Lee and H. G. Park, *Science*, 2014, **344**(6181), 289.
8. S. P. Surwade, S. N. Smirnov, I. V. Vlassiouk, R. R. Unocic, G. M. Veith, S. Dai and S. M. Mahurin, *Nat. Nanotechnol.*, 2015, **10**(5), 459.
9. S. C. O'Hern, M. S. Boutilier, J. C. Idrobo, Y. Song, J. Kong, T. Laoui, M. Atieh and R. Karnik, *Nano Lett.*, 2014, **14**(3), 1234.
10. D. E. Jiang, V. R. Cooper and S. Dai, *Nano Lett.*, 2009, **9**(12), 4019.
11. Y. Li, Z. Zhou, P. Shen and Z. Chen, *Chem. Commun.*, 2010, **46**(21), 3672.
12. S. Blankenburg, M. Bieri, R. Fasel, K. Mullen, C. A. Pignedoli and D. Passerone, *Small*, 2010, **6**(20), 2266.
13. A. W. Hauser and P. Schwerdtfeger, *Phys. Chem. Chem. Phys.*, 2012, **14**(38), 13292.
14. H. Du, J. Li, J. Zhang, G. Su, X. Li and Y. Zhao, *J. Phys. Chem. C*, 2011, **115**(47), 23261.
15. J. Schrier, *ACS Appl. Mater. Interfaces*, 2011, **3**(11), 4451.
16. J. Schrier, *ACS Appl. Mater. Interfaces*, 2012, **4**(7), 3745.
17. L. W. Drahushuk and M. S. Strano, *Langmuir*, 2012, **28**(48), 16671.
18. C. Sun, M. S. Boutilier, H. Au, P. Poesio, B. Bai, R. Karnik and N. G. Hadjiconstantinou, *Langmuir*, 2014, **30**(2), 675.
19. M. Shan, Q. Xue, N. Jing, C. Ling, T. Zhang, Z. Yan and J. Zheng, *Nanoscale*, 2012, **4**(17), 5477.
20. K. Solvik, J. A. Weaver, A. M. Brockway and J. Schrier, *J. Phys. Chem. C*, 2013, **117**(33), 17050.
21. C. Z. Sun and B. F. Bai, *Sci. Bull.*, 2017, **62**(8), 554.
22. Y. Wang, Q. Y. Yang, C. L. Zhong and J. P. Li, *Appl. Surf. Sci.*, 2017, **407**, 532.
23. B. Wen, C. Sun and B. Bai, *Phys. Chem. Chem. Phys.*, 2015, **17**(36), 23619.

24. S. P. Koenig, L. Wang, J. Pellegrino and J. S. Bunch, *Nat. Nanotechnol.*, 2012, 7(11), 728.
25. L. Wang, L. W. Drahushuk, L. Cantley, S. P. Koenig, X. Liu, J. Pellegrino, M. S. Strano and J. S. Bunch, *Nat. Nanotechnol.*, 2015, **10**(9), 785.
26. L. W. Drahushuk, L. Wang, S. P. Koenig, J. S. Bunch and M. S. Strano, *ACS Nano*, 2016, **10**(1), 786.
27. M. S. Boutilier, C. Z. Sun, S. C. O'Hern, H. Au, N. G. Hadjiconstantinou and R. Karnik, *ACS Nano*, 2014, **8**(1), 841.
28. M. S. Boutilier, N. G. Hadjiconstantinou and R. Karnik, *Nanotechnology*, 2017, **28**(18), 184003.
29. M. S. Boutilier, D. Jang, J. C. Idrobo, P. R. Kidambi, N. G. Hadjiconstantinou and R. Karnik, *ACS Nano*, 2017, **11**(6), 5726.
30. R. A. Sampson, *Philos. Trans. R. Soc., London A*, 1891, **182**, 449.
31. K.-K. Tio and S. S. Sadhal, *Appl. Sci. Res.*, 1994, **52**, 1.
32. R. Zan, Q. M. Ramasse, U. Bangert and K. S. Novoselov, *Nano Lett.*, 2012, **12**(8), 3936.
33. J. Lee, Z. Yang, W. Zhou, S. J. Pennycook, S. T. Pantelides and M. F. Chisholm, *Proc. Natl. Acad. Sci. U. S. A.*, 2014, **111**(21), 7522.
34. K. He, A. W. Robertson, C. Gong, C. S. Allen, Q. Xu, H. Zandbergen, J. C. Grossman, A. I. Kirkland and J. H. Warner, *Nanoscale*, 2015, 7(27), 11602.
35. M. E. Suk and N. R. Aluru, *RSC Adv.*, 2013, **3**(24), 9365.
36. J. K. Holt, H. G. Park, Y. Wang, M. Stadermann, A. B. Artyukhin, C. P. Grigoropoulos, A. Noy and O. Bakajin, *Science*, 2006, **312**(5776), 1034.
37. I. Hanasaki and A. Nakatani, *J. Chem. Phys.*, 2006, **124**(14), 144708.
38. Z. Qin and M. J. Buehler, *Nano Lett.*, 2015, **15**(6), 3939.
39. D. Cohen-Tanugi and J. C. Grossman, *Nano Lett.*, 2012, **12**(7), 3602.
40. D. Cohen-Tanugi and J. C. Grossman, *J. Chem. Phys.*, 2014, **141**(7), 074704.
41. K. Sint, B. Wang and P. Král, *J. Am. Chem. Soc.*, 2008, **130**(49), 16448.
42. S. Zhao, J. Xue and W. Kang, *J. Chem. Phys.*, 2013, **139**(11), 114702.
43. M. E. Suk and N. R. Aluru, *J. Chem. Phys.*, 2014, **140**(8), 084707.
44. M. E. Suk and N. R. Aluru, *J. Phys. Chem. Lett.*, 2010, **1**(10), 1590.
45. Z. He, J. Zhou, X. Lu and B. Corry, *ACS Nano*, 2013, 7(11), 10148.
46. M. Heiranian, A. B. Farimani and N. R. Aluru, *Nat. Commun.*, 2015, **6**, 8616.
47. S. Garaj, W. Hubbard, A. Reina, J. Kong, D. Branton and J. A. Golovchenko, *Nature*, 2010, **467**(7312), 190.
48. T. Jain, B. C. Rasera, R. J. Guerrero, M. S. Boutilier, S. C. O'Hern, J. C. Idrobo and R. Karnik, *Nat. Nanotechnol.*, 2015, **10**(12), 1053.
49. J. Feng, K. Liu, M. Graf, D. Dumcenco, A. Kis, M. Di Ventra and A. Radenovic, *Nat. Mater.*, 2016, **15**(8), 850.
50. J. Feng, M. Graf, K. Liu, D. Ovchinnikov, D. Dumcenco, M. Heiranian, V. Nandigana, N. R. Aluru, A. Kis and A. Radenovic, *Nature*, 2016, **536**(7615), 197.
51. A. Gadaleta, C. Sempere, S. Gravelle, A. Siria, R. Fulcrand, C. Ybert and L. Bocquet, *Phys. Fluids*, 2014, **26**(1), 012005.

52. R. C. Rollings, A. T. Kuan and J. A. Golovchenko, *Nat. Commun.*, 2016, 7, 11408.
53. S. C. O'Hern, C. A. Stewart, M. S. Boutilier, J. C. Idrobo, S. Bhaviripudi, S. K. Das, J. Kong, T. Laoui, M. Atieh and R. Karnik, *ACS Nano*, 2012, **6**(11), 10130.
54. S. C. O'Hern, D. Jang, S. Bose, J. C. Idrobo, Y. Song, T. Laoui, J. Kong and R. Karnik, *Nano Lett.*, 2015, **15**(5), 3254.
55. P. R. Kidambi, R. A. Terry, L. Wang, M. S. H. Boutilier, D. Jang, J. Kong and R. Karnik, *Nanoscale*, 2017, **9**(24), 8496.
56. M. Miao, M. B. Nardelli, Q. Wang and Y. Liu, *Phys. Chem. Chem. Phys.*, 2013, **15**(38), 16132.
57. S. Hu, M. Lozada-Hidalgo, F. C. Wang, A. Mishchenko, F. Schedin, R. R. Nair, E. W. Hill, D. W. Boukhvalov, M. I. Katsnelson, R. A. Dryfe, I. V. Grigorieva, H. A. Wu and A. K. Geim, *Nature*, 2014, **516**(7530), 227.
58. J. L. Achtyl, R. R. Unocic, L. Xu, Y. Cai, M. Raju, W. Zhang, R. L. Sacci, I. V. Vlassiouk, P. F. Fulvio, P. Ganesh, D. J. Wesolowski, S. Dai, A. C. van Duin, M. Neurock and F. M. Geiger, *Nat. Commun.*, 2015, **6**, 6539.
59. M. I. Walker, P. Braeuninger-Weimer, R. S. Weatherup, S. Hofmann and U. F. Keyser, *Appl. Phys. Lett.*, 2015, **107**(21), 213104.
60. I. Poltavsky, L. Zheng, M. Mortazavi and A. Tkatchenko, *Arxiv*, 2016.
61. M. Seel and R. Pandey, *2D Mater.*, 2016, **3**(2), 025004.
62. Q. Zhang, M. Ju, L. Chen and X. C. Zeng, *J. Phys. Chem. Lett.*, 2016, 7(17), 3395.
63. M. Lozada-Hidalgo, S. Hu, O. Marshall, A. Mishchenko, A. N. Grigorenko, R. A. Dryfe, B. Radha, I. V. Grigorieva and A. K. Geim, *Science*, 2016, **351**(6268), 68.
64. L. Shi, A. Xu, G. Chen and T. Zhao, *J. Phys. Chem. Lett.*, 2017, **8**(18), 4354.
65. J. M. Kroes, A. Fasolino and M. I. Katsnelson, *Phys. Chem. Chem. Phys.*, 2017, **19**(8), 5813.
66. L. Tsetseris and S. T. Pantelides, *Carbon*, 2014, **67**, 58.
67. M. Lozada-Hidalgo, S. Zhang, S. Hu, A. Esfandiar, I. V. Grigorieva and A. K. Geim, *Nat. Commun.*, 2017, **8**, 15251.
68. B. Radha, A. Esfandiar, F. C. Wang, A. P. Rooney, K. Gopinadhan, A. Keerthi, A. Mishchenko, A. Janardanan, P. Blake, L. Fumagalli, M. Lozada-Hidalgo, S. Garaj, S. J. Haigh, I. V. Grigorieva, H. A. Wu and A. K. Geim, *Nature*, 2016, **538**(7624), 222.
69. L. Wang, C. M. Williams, M. S. H. Boutilier, P. R. Kidambi and R. Karnik, *Nano Lett.*, 2017, **17**(5), 3081.
70. C. Lee, X. Wei, J. W. Kysar and J. Hone, *Science*, 2008, **321**(5887), 385.
71. D. Cohen-Tanugi and J. C. Grossman, *Nano Lett.*, 2014, **14**(11), 6171.

CHAPTER 4

Selective Mass Transport Properties of Graphene Oxide Lamellar Membranes

PENGZHAN SUN*[a] AND HONGWEI ZHU[b]

[a] School of Physics and Astronomy, University of Manchester, Manchester M13 9PL, UK; [b] State Key Laboratory of New Ceramics and Fine Processing, School of Materials Science and Engineering, Tsinghua University, Beijing 100084, China
*Email: pengzhan.sun@manchester.ac.uk

4.1 Transport Properties of Water Through GO Membranes

Nair, *et al.*[1] first investigated the mass transport properties of GO membranes. They prepared GO membranes by spray- or spin-coating. The as-prepared membranes were freestanding with a lamellar structure. They proposed that the sp^2 nanocapillary network formed in the interlayer galleries between adjacent GO nanosheets was responsible for mass transport through GO membranes, as illustrated in Figure 4.1a. After mounting the as-prepared freestanding GO membrane into a self-made container, the mass transport properties were studied. They showed that the micron- or submicron-thick GO membranes were impermeable to the smallest gaseous atoms, *i.e.* helium. Similar phenomena also occurred in the cases of organic liquids and vapors such as hexane and ethanol. Unexpectedly, water vapors could permeate through GO membranes unimpededly, in nearly the same

Nanoscience & Nanotechnology Series No. 47
Graphene-based Membranes for Mass Transport Applications
Edited by Pengzhan Sun and Hongwei Zhu

Published by the Royal Society of Chemistry, www.rsc.org

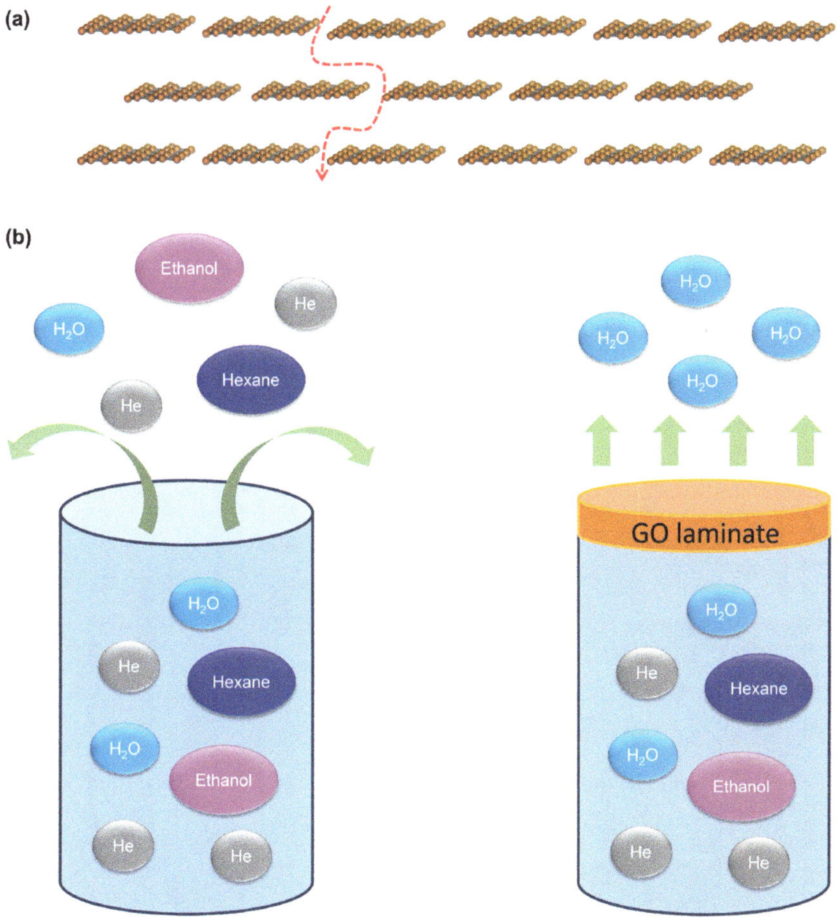

Figure 4.1 (a) Structural illustration of as-prepared GO membrane. (b) Schematic drawings of the mass transport properties of GO membranes.

way as through an open aperture. These results are schematically drawn in Figure 4.1b. By comparing the transmembrane permeabilities of different substances, it was concluded that the as-prepared GO membranes were impermeable to all the liquids, vapors and gases (including the smallest helium), but allowed the unimpeded permeation of water vapors. These seemingly incompatible results were attributed to the low friction flow of ordered water layers through the nanocapillary network within GO membranes.

The above work unveiled the unique unimpeded permeation of water vapor through GO membranes. However, the transmembrane transport properties of liquid water without external pressure, which was found to be a much more complex system, remained unknown, because of the lack of macro- or microscopic differences between the feed and permeate water separated by the central GO membrane. In order to overcome this obstacle, Sun, *et al.*[2,3]

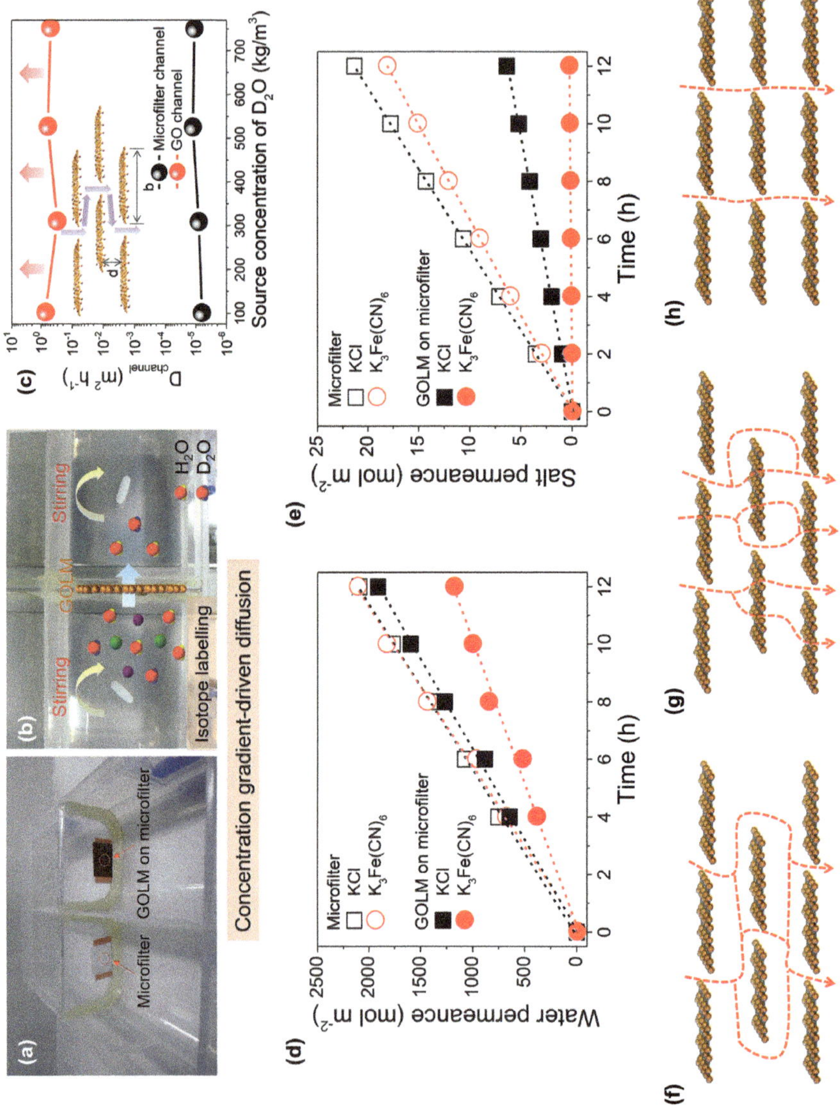
(a)
Microfilter
GOLM on microfilter
(b)
Stirring
GOLM
Stirring
H2O
D2O
Isotope labelling
Concentration gradient-driven diffusion
(c)
Microfilter channel
GO channel
Source concentration of D2O (kg/m3)
(d)
Water permeance (mol m-2)
Time (h)
Microfilter
KCl
K3Fe(CN)6
GOLM on microfilter
KCl
K3Fe(CN)6
(e)
Salt permeance (mol m-2)
Time (h)
(f)
(g)
(h)

developed an isotope labelling technique. By dissolving a certain amount of D_2O (an isotope of H_2O) in the feed water, the transmembrane permeation of D_2O was monitored instead. By quantifying the transmembrane permeation differences between D_2O and H_2O, the transport properties of H_2O were extrapolated. Employing this isotope labelling technique, the permeation properties of liquid water as well as solvent water in the presence of dissolved molecules or ions could be investigated. The GO membranes in their cases were prepared by vacuum-filtration with the assistance of PVDF microfilters (pore diameter: ~0.2 μm, porosity: ~80%). After being sealed into a self-made permeation apparatus, which contained a leak hole separating the feed reservoir (filled with D_2O-labelled water or salt solution) from the permeate reservoir (filled with pure water) (Figure. 4.2a, b), a series of mass transport experiments were conducted. By performing identical experiments with bare PVDF microfilters as the separation membranes, the effect of the polymeric support on the permeation properties of GO membranes could be subtracted. If the nanochannels formed within the interlayer spacing of GO membranes were considered to be the main transport pathways, which seemed to be reasonable because of the presence of only a tiny areal ratio of intrinsic defects within GO flakes that could be easily blocked by the adjacent sp^3 oxidized regions upon mutually stacking,[4–6] as illustrated in the inset of Figure 4.2c, the diffusivities of liquid water labelled with various concentrations of D_2O through the gallery channels of GO membranes could be evaluated, in comparison with those through the submicron-sized polymeric channels within microfilters, as shown in Figure 4.2c. The diffusivities of liquid water through the interlayer nanochannels within GO membranes were estimated to be ~5 orders of magnitude greater than those through the submicron-sized channels within microfilters (*i.e.* the bulk diffusion case), in exact agreement with the results obtained with water vapors,[1] showing the ultrafast permeation of liquid water through the interlayer nanochannels in GO membranes. Figure 4.2d and e show the representative water and salt permeances of D_2O (30 wt%)-labelled KCl and $K_3Fe(CN)_6$ solutions (0.1 M) through bare PVDF microfilters and GO membranes, respectively. The water permeations were identical for both KCl and $K_3Fe(CN)_6$ feed solutions through microfilters (Figure 4.2d), whereas the salt

Figure 4.2 (a, b) Experimental setups for isotope labelling-assisted investigation of liquid water transmembrane permeation. (c) Diffusivities of liquid water through the assumed interlayer nanochannels within GO membranes (inset) and the submicron-sized polymeric channels in the microfilter, respectively, as a function of D_2O feed concentration. Reproduced from ref. 2 with permission from the Royal Society of Chemistry. (d, e) Water and salt permeations through microfilters and GO membranes on microfilters, respectively, for representative KCl and $K_3Fe(CN)_6$ solutions. Feed concentration of salt: 0.1 M, feed concentration of D_2O: 30 wt%. (f–h) Assumed mass transport pathways in GO laminates. Reproduced from ref. 3 with permission from the Royal Society of Chemistry.

permeation of $K_3Fe(CN)_6$ was slightly slower than that of KCl (Figure 4.2e). With ~2.5 µm-thick GO membrane deposited on the microfilter, the water permeation of KCl feed solution was reduced by just 8% compared to the bare microfilter case, whereas that of the $K_3Fe(CN)_6$ feed solution was reduced by 44%, presumably attributed to the blockage of water transport by the larger sized $K_3Fe(CN)_6$. By contrast, the salt permeations of KCl and $K_3Fe(CN)_6$ through GO membrane on the microfilter were reduced by 70% and 98%, respectively, compared to those through the bare microfilter. These results indicated that solvent water could permeate through GO membranes rapidly except when partially blocked by larger solutes such as $K_3Fe(CN)_6$, whereas salts could be effectively captured or blocked by GO membranes due to some kinds of salt–GO interactions or steric hindrance, which will be discussed in detail in the next section.

The diffusivity values (~5 orders of magnitude greater than the bulk case) through the nanochannels in GO membranes were estimated based on the model that mass transport mainly occurred through the tortuous nanocapillaries formed in the interlayer galleries between adjacent GO layers (Figure 4.2f), from which the conclusion of ultrafast liquid water permeation through GO nanochannels was drawn. However, in the as-prepared GO lamellar membranes, not only the interlayer galleries, but also the point defects (*e.g.* nanopores) within GO sheets and the inter-edge channels between GO sheets in the same layers might act as mass transport pathways, which would significantly affect the conclusions. According to the high-resolution transmission electron microscopy (HRTEM) studies of GO nanosheets,[4-6] the areal ratios of the nanopore, sp^2 pristine graphitic and sp^3 oxidized regions were approximately 2%, 16% and 82%, respectively. Such a tiny areal percentage of nanopores on GO nanosheets resulted in their easy blockage by the adjacent wide-extended impermeable sp^3 regions. Therefore, the leakage through nanopores on GO flakes could be neglected. Assuming a critical case[3] that the nanopores were completely unblocked by the oxidized regions on adjacent GO layers and acted as extra percolation pathways for reducing the length of mass transport nanochannels (Figure 4.2g), the estimated water diffusivities were roughly 2 orders of magnitude lower than those estimated from Figure 4.2f. Assuming another critical case[3] that water molecules directly transported through the perpendicular pathways formed by directly connecting the inter-edge slits without any turns through the nanosheet surfaces (Figure 4.2h), which significantly reduced the length of nanochannels by approximately 3 orders of magnitude, then the water diffusivities were estimated to be 3 orders of magnitude lower than those in Figure 4.2f. Despite these two unrealizable critical cases, the estimated water diffusivities through GO nanochannels were still orders of magnitude greater than the bulk case, indicating that liquid water could indeed permeate through GO membranes rapidly.

A simulation study was employed to explore the origin of the anomalous ultrafast water permeation through GO membranes. One representative first-principles calculation[7] is shown in Figure 4.3a. The migration of

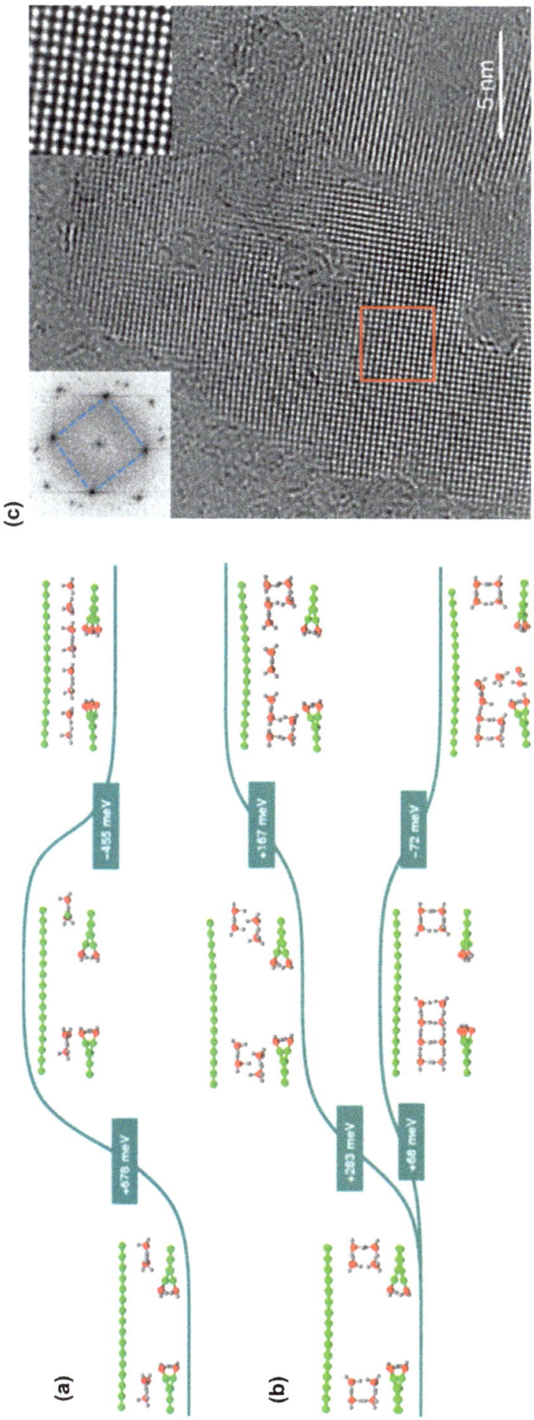

Figure 4.3 Energy costs and optimistic atomic structures for (a) monolayer water migrating through hydroxyl terminated graphene pore edge and (b) bilayer water migrating with respect to each other and through hydroxyl terminated graphene pore edge, respectively. Reprinted with permission from D. W. Boukhvalov, M. I. Katsnelson and Y. W. Son, *Nano Lett.*, 2013, **13**, 3930, Copyright 2013 American Chemical Society.[7] (c) HRTEM image of water molecules confined in graphene capillary. The insets show Fourier transform of the entire area (top left) and magnified view of the region outlined in red (top right), respectively. Reprinted by permission from Macmillan Publishers Ltd: *Nature* (G. Algara-Siller, O. Lehtinen, F. C. Wang, R. R. Nair, U. Kaiser, H. A. Wu, A. K. Geim and I. V. Grigorieva, *Nature*, 2015, **519**, 443) Copyright 2015, Nature Publishing Group.[8]

ordered water layers with respect to a hydroxyl-terminated GO sheet in the monolayer case and that with respect to an adjacent water layer in the bilayer case both gave rise to a significant rise in energy barrier, and thus they were energetically unfavorable. By contrast, calculations show that in the case of bilayer water, a mechanism of the collapse of ordered water layers and further collective drop to the next interlayer void where they formed another bilayer water was the most energetically favorable. According to XRD characterizations on dry and fully wet GO membranes,[3] significant swelling would occur in a water environment. The enlarged interlayer galleries in GO membranes might serve as a certain kind of template for the nucleation and growth of ordered water multilayers. Therefore, the above calculation results might provide a reasonable explanation for the origin of ultrafast water permeation through GO membranes. Recently, a HRTEM study[8] focusing on the structure of water locked in between two pristine graphene sheets directly observed the formation of square ice in graphene nanocapillaries (Figure 4.3b), which might experimentally shed light on the possible mechanism of ultrafast water permeation through the sp^2 nanocapillaries within GO membranes, that is, the formation of ordered water layers and their collective low-friction flow in sp^2 nanocapillaries might give rise to the anomalous water transport behavior in GO membranes.

On the other hand, several more recent simulation studies[9-11] showed the breakdown of ultrafast water transport in GO membranes due to the side-pinning effect on the nanoconfined water flows originating from the oxygen functionalities around sp^2 nanocapillaries. The permeability through GO-based membranes could be quantitatively predicted by a continuum hydrodynamics model, by properly considering the entrance, slippage and viscosity of the nanoconfined water flow[9-12]. On the basis of the above calculations, it is more reasonable to assume that the fast water flow in GO membranes might arise from their porous and defective microstructures (*e.g.* substantial percolations through wide channels formed at wrinkles, pinholes and inter-edge slits). According to the estimations of water diffusivities[3] in Figure 4.2f–g, if water flow in GO membranes was assumed to be direct through the quasi-vertically aligned pathways formed by nanopores on GO sheets and inter-edge spaces between GO sheets without many tortuous turns across the nanosheet surfaces, the estimated water diffusivities in GO nanochannels would approach the bulk case. Despite this model being rather ideal without considering the boundary conditions of possible barriers to the water flow at the sheet edges, it provided a reasonable direction for quantitatively exploring the water transport mechanism through GO-based laminates. Up to now, there still remains a huge space for investigating the mechanism related to fast water flow in GO layers and further efforts on this aspect are still needed.

4.2 Transport Properties of Dissolved Solutes in Water Through GO Membranes

In regard to the transmembrane transport properties of solutes dissolved in water, Joshi *et al.*[13] prepared a micrometer-thick GO membrane *via* vacuum filtration, placed it into a U-shaped tube to separate the feed and permeate reservoirs, and investigated the transport properties of a wide range of solute molecules and ions dissolved in water. In the dry state, the as-prepared GO membranes were vacuum-tight and impermeable to even the smallest gas atoms (helium). On the other hand, after immersion in water, these membranes could act as precise molecular sieves to block all the solute molecules and ions with hydrated dimeters of >0.9 nm, as schematically drawn in Figure 4.4. Smaller solute ions could permeate through GO membranes with a rate orders of magnitude faster than the prediction from simple diffusion. The excellent molecular sieving effect of GO membranes was attributed to the steric hindrance effect of sp^2 nanocapillaries that opened up in water to accept the species with fitted dimensions, while the fast transport of small ions originated from the nanoconfinement-induced high capillary pressure that acted on ions in nanocapillaries.

The precise size-dependent selectivity of mass transport through GO membranes endows them with promise in filtration and separation. Inspired by these results, Mi[14] proposed that by adjusting the dimension of

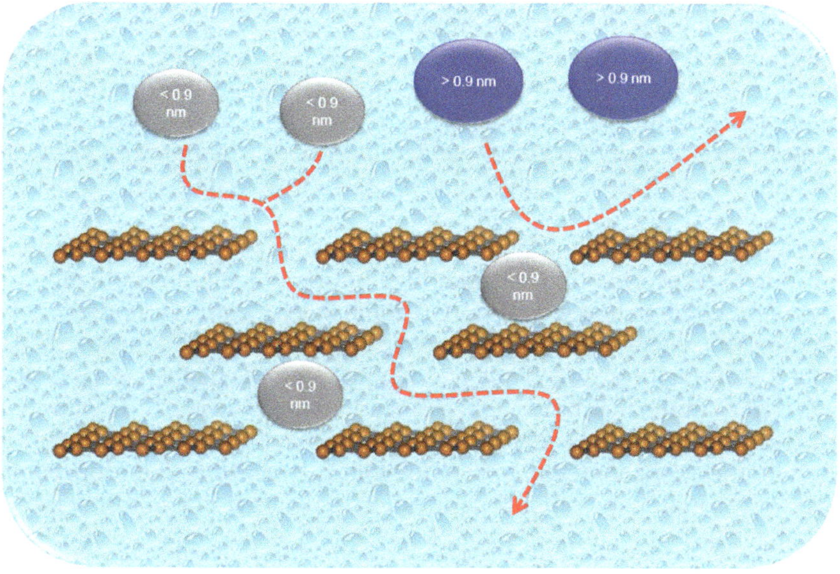

Figure 4.4 Schematic diagram illustrating the molecular sieving effect of GO membrane toward different sized solutes in water.

nanocapillaries formed in the interlayer galleries of GO membranes, multifunctional separation applications could be achieved. Briefly, by adopting appropriate physical or chemical routes to partially reduce the as-prepared GO membranes for a narrowed interlayer spacing (<0.7 nm), inserting selected chemical groups or polymer chains for a moderate interlayer spacing ($1 \sim 2$ nm) or intercalating larger sized 0D-nanoparticles or 1D-nanofibers for an enlarged interlayer spacing (>2 nm), excellent water desalination, wastewater purification and reuse as well as pharmaceutical and fuel separation, and artificial kidneys and dialysis might be achieved, respectively. These proposals show the excellent flexibility of GO-based membranes for multifunctional filtration and separation applications, presumably readily realized by tuning their interlayer gallery spacing.

Sun, *et al.*[15] systematically investigated the transmembrane transport properties of dissolved small hydrated ions in water. They fabricated a freestanding GO membrane by a simple drop-casting method, sealed it into a self-made permeation apparatus (shown in Figure 4.2a, b) to separate the feed reservoir from the permeate, injected equivalent volumes of a certain salt solution (0.1 M) and deionized water into the feed and permeate reservoirs, respectively, and monitored the transmembrane permeances of feed ions in real time. As shown in Figure 4.5a, during the process of feed ions permeating through GO membranes, excellent selectivity could be achieved. Sodium salts permeated through GO membranes rapidly, whereas transition metal salts permeated through much slower. It was interesting to find that the copper salts were almost entirely blocked by GO membranes. As shown in Figure 4.5b, the permeation properties of organic dyes (represented by RhB molecules) through GO membranes were also tested for comparison, revealing that after 3 h of permeation, the RhB molecules were also completely blocked by GO membranes. These results indicate that excellent permeation selectivity could be achieved with GO membranes, showing potential in environmental protection areas such as waste water purification and reuse.

Next, by employing natural graphite and worm-like expanded graphite as the source, respectively, in the modified Hummers' method, micron- and several hundred nanometer-sized GO nanosheets were synthesized. Freestanding GO membranes were prepared by drop-casting aqueous dispersions containing these two kinds of different-sized GO nanosheets and were subjected to transmembrane ion permeation experiments[16]. As shown in Figure 4.5c and d, increasing the lateral dimension of GO nanosheets from several hundred nanometers to several micrometers led to the proportional decrease of ion transmembrane permeation, indicating that tailoring the GO building blocks in lamellar membranes could effectively tune the mass transport properties of GO membranes by engineering their percolating nanochannels. Specifically, as shown in Figure 4.5c, for main-group cations, excellent selectivity could also be achieved with GO membranes, with the through-membrane permeating rate order as follows: $H^+ > Na^+$ and $Mg^{2+} > K^+$, Ca^{2+} and Ba^{2+}.

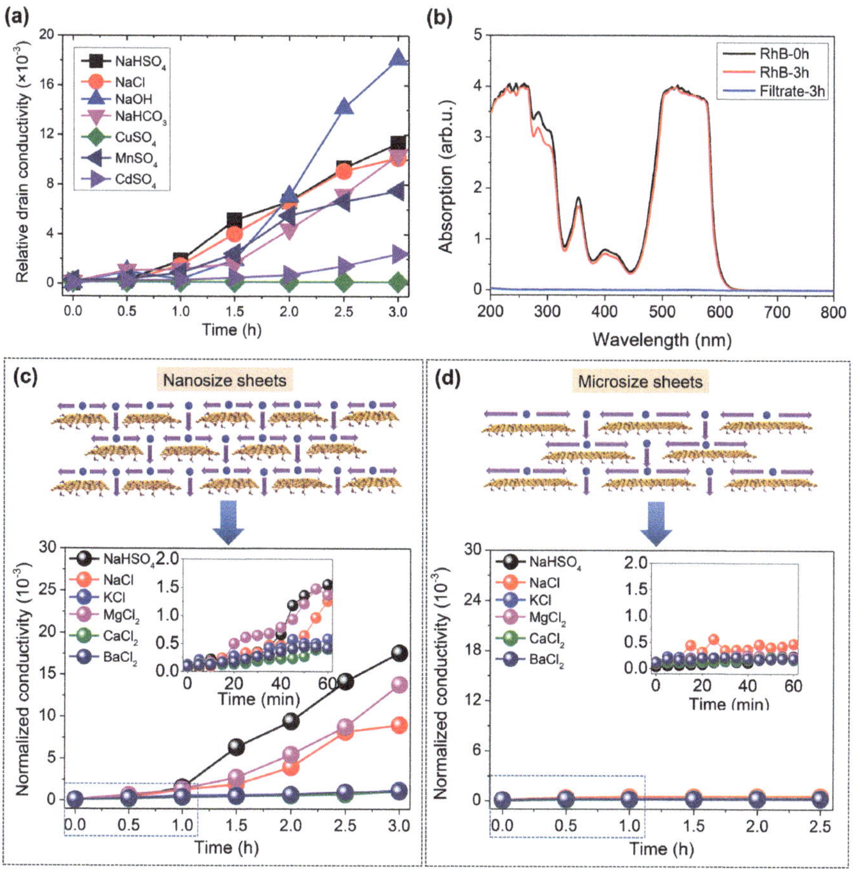

Figure 4.5 (a) Permeations of various salts through freestanding GO membranes. (b) UV-*vis* absorption spectra for RhB feed solutions before and after 3 h-transmembrane permeation process and the corresponding permeate solution. Reprinted with permission from P. Sun, M. Zhu, K. Wang, M. Zhong, J. Wei, D. Wu, Z. Xu and H. Zhu, *ACS Nano*, 2013, **7**, 428, Copyright 2013 American Chemical Society.[15] (c, d) Permeations of various salts through freestanding GO membranes composed of several hundred nanometer- and several micron-sized GO nanosheets, respectively. The top insets illustrate the corresponding permeation pathways. Reprinted with permission from P. Sun, F. Zheng, M. Zhu, Z. Song, K. Wang, M. Zhong, D. Wu, R. B. Little, Z. Xu and H. Zhu, *ACS Nano*, 2014, **8**, 850, Copyright 2014 American Chemical Society.[16]

Based on the above results, the origin of the selective ion permeation mechanism of GO membranes was explored,[15,16] as illustrated in Figure 4.6. Due to the unique structure of enriched oxygen-containing functional groups and sp^2 aromatic nanoclusters co-existing within a 2D lattice,[17,18] various ions tend to interact with the different parts of GO *via* diverse electrostatic or chemical interactions with different strengths. In detail, transition metal cations preferred to interact with the sp^3 C–O matrix *via*

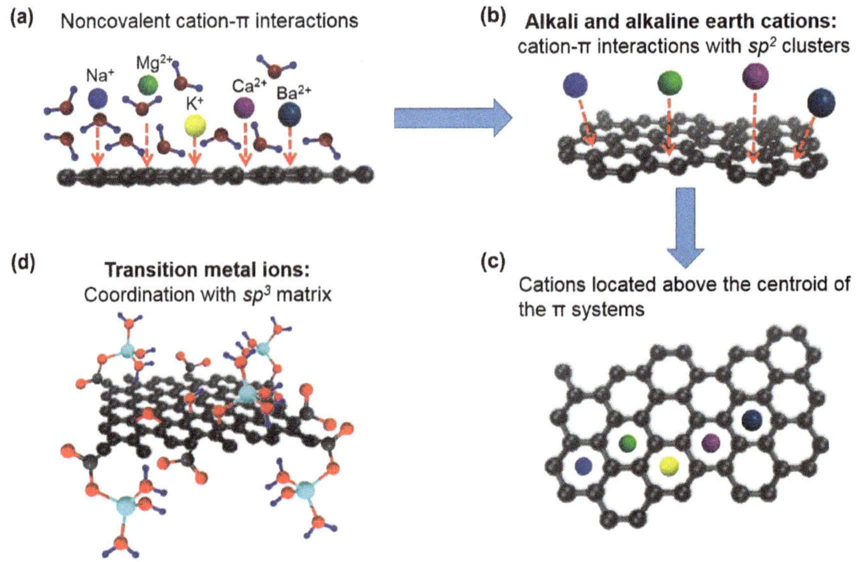

Figure 4.6 Diverse chemical interactions between ions and active sites on GO that are responsible for the selectivity of GO membranes. (a–c) Cation–π interaction of alkali and alkaline earth cations with sp^2 C–C clusters on GO. (d) Coordination interaction of transition metal cations with sp^3 C–O matrix on GO.
Reprinted with permission from P. Sun, F. Zheng, M. Zhu, Z. Song, K. Wang, M. Zhong, D. Wu, R. B. Little, Z. Xu and H. Zhu, *ACS Nano*, 2014, **8**, 850, Copyright 2014 American Chemical Society.[16]

coordination interaction.[15] Due to the lack of *d* or *f* electrons, alkali and alkaline earth cations tended to interact with the sp^2 aromatic C–C clusters *via* noncovalent cation–π interaction.[16] In the aqueous environment, the deionization of oxygen functionalities attached on GO would charge the membrane negatively, and anions tended to be repulsed electrostatically, resulting in a passive transmembrane permeation process. Due to the electroneutrality requirement of the permeate solutions, anions would be dragged by the forwardly permeating cations through the membrane and finally to the permeate side.[19] Therefore, the balance between the diverse ion–GO interactions and the ion hydration–dehydration in confined GO nanochannels led to the excellent selectivity of GO membranes when different ions permeated through. The above proposed mechanism has been confirmed experimentally (*e.g.* by Auger Electron Spectroscopy (AES) and X-ray Photoelectron Spectroscopy (XPS) characterizations of post-permeated GO membranes,[15,16] and ions permeating through different functionalized graphene membranes[20]) and theoretically (*e.g.* by first-principles calculations[16]).

To this end, the unique transmembrane mass transport properties of GO membranes have been overviewed by especially emphasizing the permeation behaviors of solvent water and solute ions and molecules. This revealed that water molecules (either in vapor or liquid forms) could demonstrate ultrafast permeation through GO membranes,[1-3] whereas a wide series of dissolved molecules and ions could be effectively separated based on steric hindrance effects from sp^2 nanocapillaries[13] and diverse electrostatic or chemical interactions from different regions of GO.[15,16,19,20] These results inspired research into whether there existed significant selectivity between the solvent water and different solutes dissolved, which would provide several meaningful developing directions involving GO membranes ranging from water desalination to saline solution concentration. Accordingly, by employing the isotope labelling technique proposed above, Sun, *et al.*[3] measured the concentration gradient-driven diffusions of solvent water and solute salt ions through GO membranes simultaneously, based on which, the diffusivities of water and salts were calculated. As shown in Figure 4.7a, the diffusivities of solvent water were all significantly greater than those of salts, indicating that salt ions could be effectively removed with respect to water, showing promise in water desalination. Among various salts, excellent selectivity in transmembrane diffusivity was achieved, due to the steric hindrance from sp^2 nanocapillaries and diverse electrostatic and chemical interactions from different regions of GO, in good agreement with the mechanism proposed in Figure 4.6. Notably, the transmembrane diffusivities of solvent water were lower than that of pure water and different saline solutions gave rise to different solvent water diffusivities, indicating that the confined ion–water interactions (*i.e.* ion hydration–dehydration) in GO interlayer galleries caused different degrees of blockage to the corresponding solvent water transport.

In order to further quantify the relative removal of salt ions with respect to water molecules, a nominal rejection rate was defined as $R_n = 1 - D_{salt}/D_{water}$. Figure 4.7b shows the nominal rejections of representative salts when permeating through GO membranes, indicating that GO membranes possessed excellent water desalination ability. The nominal rejections of various salts seemed to be independent of the anion/cation valence ratios (Z^-/Z^+), indicating that the electrostatic interactions between charged ions and GO might not play a dominant role in desalination with GO membranes. This could be further confirmed by pH- and electrolyte concentration-dependent nominal rejections of GO membranes, as shown in Figure. 4.7c, d. Varying the pH values of KCl feed solution led to negligible changes in nominal rejection (Figure 4.7c), indicating that the ion exclusion of GO membranes was not sensitive to surface charge density variations. On the other hand, by plotting the nominal rejections of KCl feed solution as a function of Debye length (λ_D, $\lambda_D \propto c^{-1/2}$) in Figure 4.7d (the relationship between nominal rejection and electrolyte concentration is shown in the inset), it is evident that the nominal rejection dropped sharply at a λ_D of ~0.7 nm, which is significantly smaller than the void spacing in the interlayer galleries (*i.e.* sp^2 nanocapillaries) of GO membranes (~1 nm, obtained from previous XRD characterizations[3]), indicating that the ion exclusion of GO membranes was not sensitive to surface charge screening. The above mechanism discussions indicated that the diverse chemical interactions between ions and GO

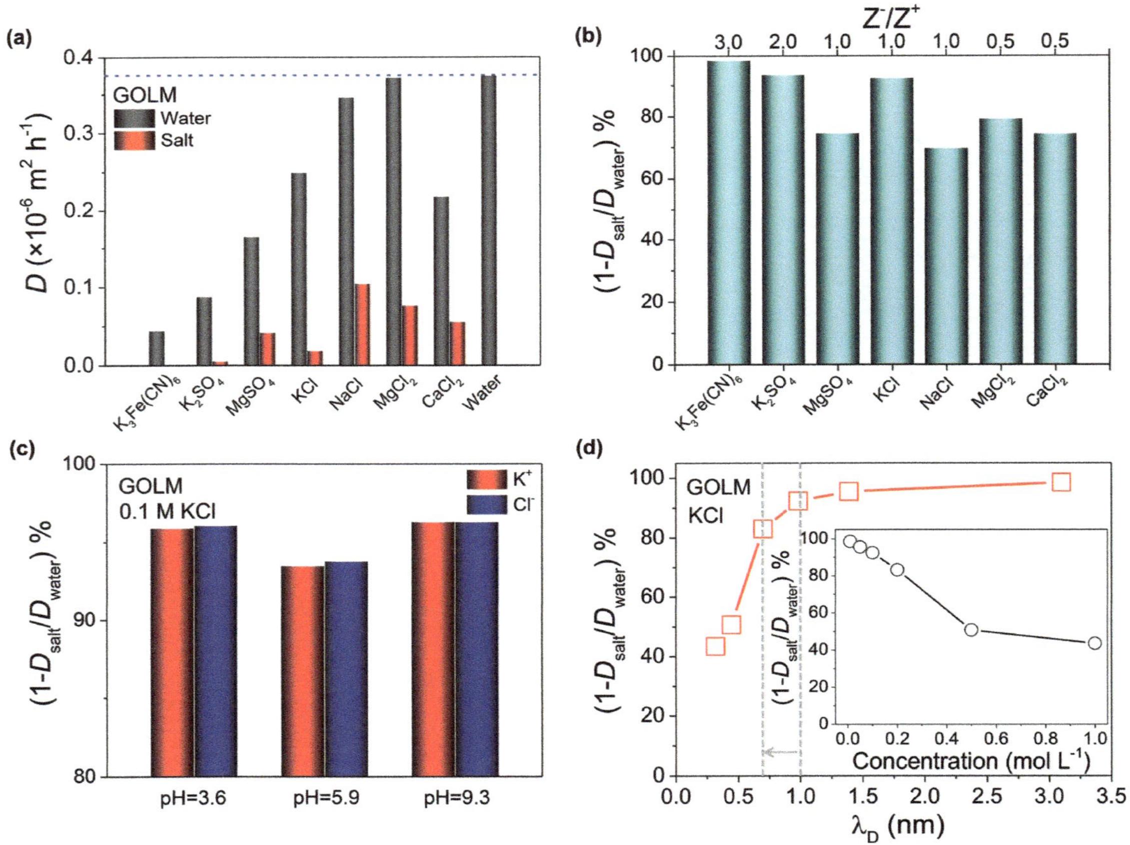

Figure 4.7 (a) Diffusivities of various salts and corresponding solvent water through GO membranes. (b) Calculated nominal rejections ($R_n = 1 - D_{salt}/D_{water}$) of various salts. (c) pH-dependent nominal rejections of K^+ and Cl^- ions. (d) Nominal rejections of KCl as a function of Debye length (λ_D). The inset shows the relationship of KCl rejections as a function of electrolyte concentration. Reproduced from ref. 3 with permission from the Royal Society of Chemistry.[3]

dominated the selective ion transport processes through GO membranes.[15,16] This suggests that the excellent ion rejection performance of GO membranes might originate from the effective anchoring or capture of ions by the diverse ion–GO chemical interactions.

To characterize the effect of nanochannel length on mass transport properties, GO membranes with gradually increased thicknesses were deposited on control microfilters through vacuum filtration and the transmembrane permeations of solvent water and selected salts were simultaneously measured by employing the well-developed isotope labelling technique,[3] as shown in Figure. 4.8a, b. It is interesting to see that with the thickening of GO membranes (*i.e.* lengthening of GO nanochannels), the transmembrane water flux remained nearly unchanged while the salt flux decreased gradually, indicating that water molecules experienced an ultrafast through-channel permeation process, independent of the nanochannel length, while salt ions were more effectively captured or anchored by the increased amount of active sites decorated around the nanochannels with their lengthening. These results were consistent with the concepts of low-friction transport of water molecules,[1,2] which gave rise to their ultrafast transmembrane permeation and diverse ion–GO interactions,[15,16,19,20] which resulted in excellent ion selectivity when permeating through. By calculating and plotting the nominal rejections as a function of GO membrane thickness in Figure 4.8c, it is evident that the rejections of GO membranes increased with their thickness, due to the invariable water flux and gradually decreased salt flux with increasing GO nanochannel length (Figure. 4.8a, b). Therefore, it is reasonable to deduce that thick GO membranes may have potential in desalination. Such unique mass transport behaviors of GO nanochannels were supported by steered molecular dynamics (SMD) simulations,[3] whereby the free energy profiles of water molecules and ions were calculated during their through-channel transport processes (Figure. 4.8d-f). A substantial energy barrier was characterized for the ion diffusion processes through GO nanochannels, which was negligible for water, in good agreement with the experimental results in Figure. 4.8a, b.

To make a brief summary, water molecules (in either vapor or liquid forms) experienced ultrafast permeation through the nanochannels within GO membranes,[1–3] whereas due to the presence of numerous sp^2 C–C clusters and various oxygen functionalities,[17,18] diverse electrostatic or chemical interactions occurred between various ions and the different regions of GO,[3,15,16,19,20] resulting in excellent ion selectivity when permeating through, as illustrated in Figure 4.9. In detail, the negative charges on GO electrostatically repulse anions and attract cations. The cations are first attracted into GO membranes, interact with the active sites on GO and permeate through, while the anions are dragged by the permeating cations, resulting in passive permeation.[19] The transition metal cations tended to interact with the sp^3 C–O matrix *via* coordination interactions,[15] while the alkali and alkaline earth cations preferred to interact with the sp^2 C–C clusters *via* cation–π interactions.[16] The balance between the diverse

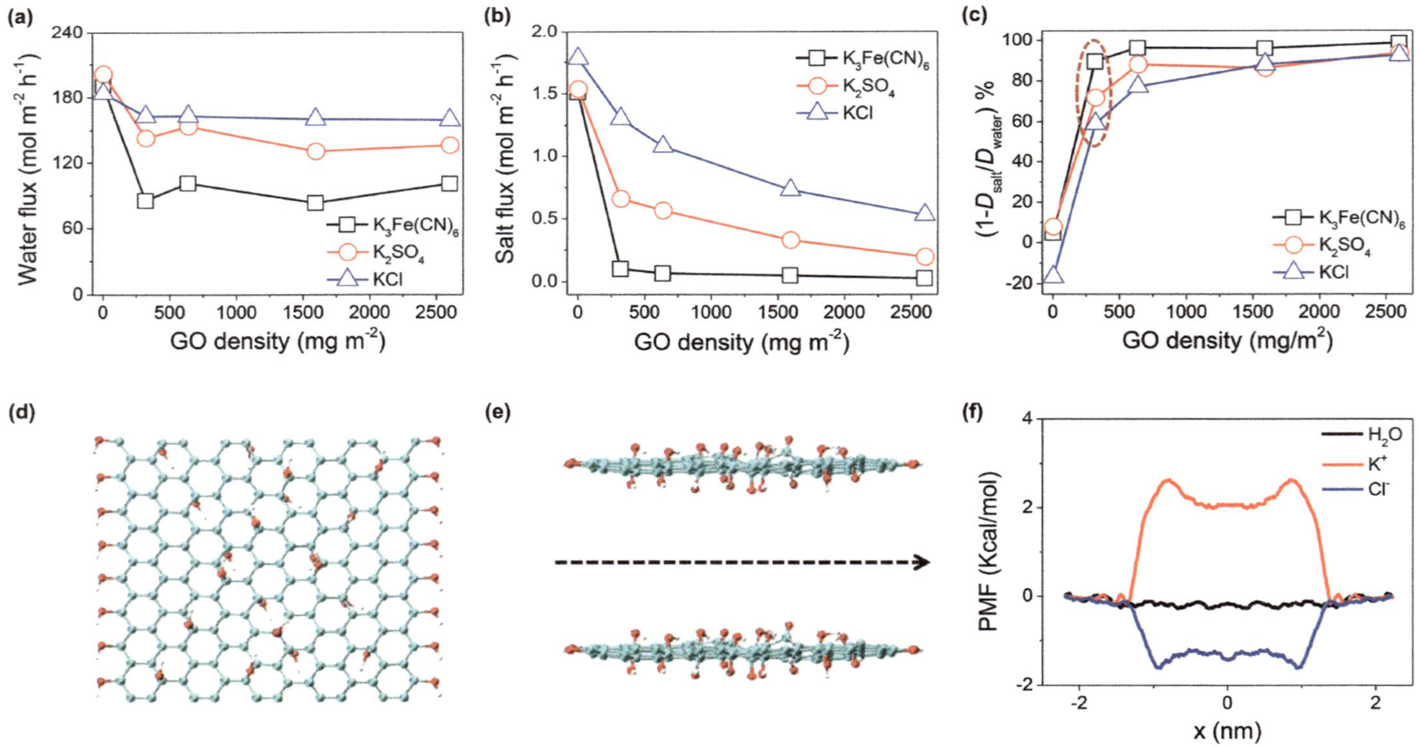

Figure 4.8 (a) Water flux, (b) salt flux and (c) corresponding nominal rejection of selected feed solutions through GO membranes (on microfilters) with increasing thickness (*i.e.* nanochannel length). Feed concentration of salts: 0.1 M, feed concentration of D_2O tracers: 30 wt%. (d, e) Atomic structures of GO nanochannels in SMD simulations. (f) Free energy profiles for the diffusions of water molecules, and K^+ and Cl^- ions through GO nanochannels in SMD simulations. Reproduced from ref. 3 with permission from the Royal Society of Chemistry.[3]

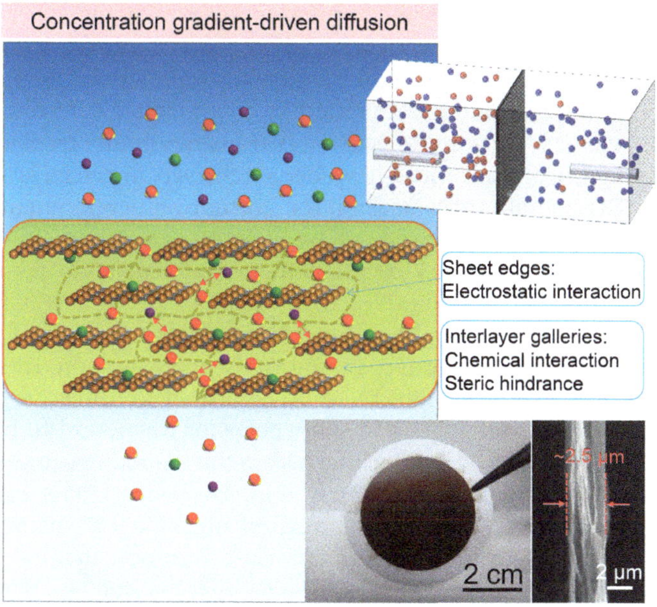

Figure 4.9 Schematic diagrams for the mechanism of water and ion transport through GO membranes and concentration gradient-driven experimental setup. Photograph of a piece of GO membrane (on PVDF microfilter) and its cross-sectional SEM image.
Reproduced from ref. 3 with permission from the Royal Society of Chemistry.[3]

electrostatic or chemical ion–GO interactions with different strengths and the ion hydration–dehydration effect confined in nanochannels gave rise to the excellent permeation selectivity of GO membranes toward different ions. Specifically, due to the different degrees of ion blockage or capture by GO arising from diverse ion–GO interactions and the ultrafast water trans-membrane permeation, intrinsic high water/ion selectivity could be achieved with GO membranes,[3] showing promise in water desalination and saline solution concentration applications.

4.3 Transport Properties of Gas Atoms and Molecules Through GO Membranes

As shown in Figure 4.1, Nair, *et al.*[1] showed that micron- or submicron-thick GO membranes were vacuum-tight and impermeable to all gas molecules and atoms. In their cases, hundreds or thousands of layers of GO nanosheets were stacked and overlapped onto each other to form a lamellar structure, in which a continuous nanocapillary network was formed for cross-membrane mass transport. The leakage through the intrinsic defects within GO nanosheets (*e.g.* pinholes or cracks) could be neglected due to their tiny areal ratio[4–6] that could easily be blocked by the adjacent impermeable oxidized matrix upon mutual stacking of numerous layers. On the other hand, by stacking only discrete layers of GO nanosheets together to form an ultrathin film without the formation of a continuous nanocapillary network, the mass transport pathways could be dominated by the intrinsic defects present on the GO nanosheets, allowing selective gas transmembrane transport. In recent years, significant advances have been achieved in selective gas separation with GO membranes. Several representative developments in this aspect[21–25] have been overviewed below.

By employing different stacking methods of graphene and GO nanosheets to prepare few-layered membranes, that is, either by contacting a porous support substrate with the air–liquid interface of a GO suspension followed by spin-coating, or by directly spin-casting a GO suspension onto the substrate surface, gas flow channels could be well engineered to achieve selective gas diffusion.[21] Such a tunable gas transport behavior was found to be strongly influenced by the degree of interlocking within the layered structure. Excellent selectivity of CO_2/N_2 was achieved with well-interlocked GO films, especially in relatively high humidity. Li *et al.*[22] prepared ultrathin GO membranes approaching 1.8 nm by a facile vacuum-assisted filtration method, and they found that a mixture separation selectivity as high as 3400 and 900 could be achieved for H_2/CO_2 and H_2/N_2, respectively. The excellent gas separation performances of the few-layered GO films obtained above were both attributed to the intrinsic structural defects within GO flakes, which acted as selective gas diffusion channels to separate target gas mixtures efficiently.

Shen *et al.*[23] assembled GO nanosheets into few layered lamellar structures and vertically inserted them into a polymer matrix by constructing hydrogen bonding between GO and polymer. The as-prepared GO-based composite membranes bearing molecular sieving interlayer galleries and straight diffusion pathways showed preferential CO_2 permeance (100 Barrier), high CO_2/N_2 selectivity (91) and excellent operational stability (>100 h). By employing an external force, GO nanosheets were oriented into lamellar membranes with precisely tunable interlayer galleries and in-plane slit-like pores.[24] The as-assembled GO laminates exhibited excellent gas molecule-sieving performances, with ~2–3 orders of magnitude higher H_2 permeance and a 3-fold enhanced H_2/CO_2 selectivity with respect to commercial gas separation membranes.

More recently, inspired by the physicochemical microenvironment in aquaporin for efficient transport of CO_2 molecules, ultrathin GO-based composite membranes with high CO_2 permeability and selectivity were prepared by vacuum filtration using borate as both a crosslinker and a transport carrier.[25] The as-prepared borate-crosslinked GO membranes showed a high CO_2 permeance and excellent selectivity toward CO_2-containing gas mixtures such as CO_2/CH_4 and CO_2/N_2. The rapid CO_2 transport was attributed to the reversible catalytic conversion of CO_2 to HCO_3^- by borate ions in the presence of water, while the efficient CO_2 separation was attributed to the precise tunability of the interlayer nanochannels by crosslinking with borate ions.

The above results indicate that excellent gas selectivity could be achieved with GO-based membranes. By finely designing the diffusion pathways within the GO lamellar structure (*e.g.* engineering the interlayer nanochannels *via* different stacking methods[21] and external force fields,[24] constructing vertically aligned diffusion pathways,[23] tailoring the percolation defects within GO flakes,[21,22] and tuning the physicochemical microenvironment of interlayer nanochannels *via* crosslinking with specific gas transport carriers[25]), desired separation performances could be obtained toward target gas mixtures. The exotic gas selectivity of GO-based membranes endows them with promise in a series of gas separation-related applications, such as CO_2 capture and H_2 separation.

4.4 Concluding Remarks

The experimental and theoretical results provided above have given a general understanding of the unique mass transport properties, related mechanisms and potential applications of GO-based membranes. These advances in this specific research area serve as a foundation for the future developments of GO-based membranes toward practical applications. Despite being promising enough, the research on this aspect is still far from mature, and it requires further efforts to be devoted to broadening the understanding of nanoscale structural characteristics, the mechanism of mass transport in nanoconfined GO channels, the through-channel transport properties of a wider range of species and the possibility of industrial-scale applications.

The existing results of water evaporation[1] and isotope labelling-assisted liquid water permeation[2,3] have shown that water molecules experienced ultrafast permeation through GO membranes, which has been attributed to the collective low-friction flow of ordered water layers through the sp^2 nanocapillaries within GO membranes. However, several simulations studies[9–12] have shown that the presence of oxygen functional groups attached around the sp^2 nanocapillaries could result in a significant side-pinning effect, which would disrupt the ordered water layers in GO nanochannels. Based on these calculations, the water transport processes in GO membranes could be predicted quantitatively through a continuum hydrodynamics model.[9–12] Therefore, serious controversies exist in the mechanism investigation of water transport through GO membranes. Notably, the experimental and theoretical results[3] displayed in Figure 4.8 show that the transmembrane water flux seemed to be independent of the GO nanochannel length with a negligible free energy barrier when permeating through, while the salt flux gradually decreased with increasing the channel length and a significant rise in energy barrier was characterized through the GO nanochannel. With lengthening the GO nanochannels, the gradual decrease of transmembrane salt flux could be easily understood because of the pinning effect from the active sites on GO, whereas the nearly constant water flux indicates that the concept of ultrafast water transmembrane transport[1] seemed to be reasonable. In short, the mechanism behind the unique water transport properties of GO membranes should be unveiled further, possibly through constructing single GO nanochannels, providing detailed structural information and *in situ* studying the flow fields inside GO nanochannels.

The mutual stacking of GO nanosheets results in a slit-like interlayer gallery. Its size and physicochemical environment can be precisely tailored by physical confinement,[26] crosslinking[25,27–30] or intercalating other guest materials with desired sizes and functions.[31–34] Abundant oxygen-containing functional groups and sp^2 aromatic C–C clusters are present on GO,[17,18] providing enriched active sites for interacting with diverse species. Therefore, by taking full advantage of the steric hindrance effect[13] and diverse electrostatic or chemical interactions[15,16,19,20] between a wide range of atoms, molecules and ions and the active sites on GO, GO-based membranes could display extraordinary permeation or transport selectivity, showing promise in a series of filtration and separation applications such as water desalination, gas separation and chemical refining. Future developments of GO-based membranes toward practical applications might include the scalable processing of thickness- and area-controlled GO-based membranes with well-defined nanochannels, improvements of membrane mechanical strength and stability (through crosslinking with polymer chains, multivalent ions and reinforcing nanomaterials) and broadening of membrane separation functions (through intercalating other 0D, 1D and 2D functional materials such as nanoparticles or nanodots, nanotubes or nanowires, metallic oxide and hydroxide nanosheets,[33–36] hBN[37,38] and transition metal chalcogenides[39–42]). Such efforts are now on the way.

References

1. R. R. Nair, H. A. Wu, P. N. Jayaram, I. V. Grigorieva and A. K. Geim, *Science*, 2012, **335**, 442.
2. P. Sun, H. Liu, K. Wang, M. Zhong, D. Wu and H. Zhu, *Chem. Commun.*, 2015, **51**, 3251.
3. P. Sun, R. Ma, H. Deng, Z. Song, Z. Zhen, K. Wang, T. Sasaki, Z. Xu and H. Zhu, *Chem. Sci.*, 2016, **7**, 6988.
4. K. Erickson, R. Erni, Z. Lee, N. Alem, W. Gannett and A. Zettl, *Adv. Mater.*, 2010, **22**, 4467.
5. D. Pacilé, J. C. Meyer, A. Fraile Rodríguez, M. Papagno, C. Gómez-Navarro, R. S. Sundaram, M. Burghard, K. Kern, C. Carbone and U. Kaiser, *Carbon*, 2011, **49**, 966.
6. C. Gomez-Navarro, J. C. Meyer, R. S. Sundaram, A. Chuvilin, S. Kurasch, M. Burghard, K. Kern and U. Kaiser, *Nano Lett.*, 2010, **10**, 1144.
7. D. W. Boukhvalov, M. I. Katsnelson and Y. W. Son, *Nano Lett.*, 2013, **13**, 3930.
8. G. Algara-Siller, O. Lehtinen, F. C. Wang, R. R. Nair, U. Kaiser, H. A. Wu, A. K. Geim and I. V. Grigorieva, *Nature*, 2015, **519**, 443.
9. N. Wei, X. Peng and Z. Xu, *Phys. Rev. E*, 2014, **89**, 012113.
10. N. Wei, X. Peng and Z. Xu, *ACS Appl. Mater. Interfaces*, 2014, **6**, 5877.

11. A. Montessori and C. A. Amadei, arXiv:1608.02098v1.
12. H. Yoshida and L. Bocquet, *J. Chem. Phys.*, 2016, **144**, 234701.
13. R. K. Joshi, P. Carbone, F. C. Wang, V. G. Kravets, Y. Su, I. V. Grigorieva, H. A. Wu, A. K. Geim and R. R. Nair, *Science*, 2014, **343**, 752.
14. B. Mi, *Science*, 2014, **343**, 740.
15. P. Sun, M. Zhu, K. Wang, M. Zhong, J. Wei, D. Wu, Z. Xu and H. Zhu, *ACS Nano*, 2013, 7, 428.
16. P. Sun, F. Zheng, M. Zhu, Z. Song, K. Wang, M. Zhong, D. Wu, R. B. Little, Z. Xu and H. Zhu, *ACS Nano*, 2014, **8**, 850.
17. G. Eda and M. Chhowalla, *Adv. Mater.*, 2010, **22**, 2392.
18. K. P. Loh, Q. Bao, G. Eda and M. Chhowalla, *Nat. Chem.*, 2010, **2**, 1015.
19. P. Sun, F. Zheng, M. Zhu, K. Wang, M. Zhong, D. Wu and H. Zhu, *Sci. Rep.*, 2014, **4**, 5528.
20. P. Sun, H. Liu, K. Wang, M. Zhong, D. Wu and H. Zhu, *J. Phys. Chem. C*, 2014, **118**, 19396.
21. H. W. Kim, H. W. Yoon, S. M. Yoon, B. M. Yoo, B. K. Ahn, Y. H. Cho, H. J. Shin, H. Yang, U. Paik, S. Kwon, J. Y. Choi and H. B. Park, *Science*, 2013, **342**, 91.
22. H. Li, Z. Song, X. Zhang, Y. Huang, S. Li, Y. Mao, H. J. Ploehn, Y. Bao and M. Yu, *Science*, 2013, **342**, 95.
23. J. Shen, G. Liu, K. Huang, W. Jin, K.-R. Lee and N. Xu, *Angew. Chem.*, 2015, **127**, 588.
24. J. Shen, G. Liu, K. Huang, Z. Chu, W. Jin and N. Xu, *ACS Nano*, 2016, **10**, 3398.
25. S. Wang, Y. Wu, N. Zhang, G. He, Q. Xin, X. Wu, H. Wu, X. Cao, M. D. Guiver and Z. Jiang, *Energy Environ. Sci.*, 2016, **9**, 3107.
26. J. Abraham, K. S. Vasu, C. D. Williams, K. Gopinadhan, Y. Su, C. T. Cherian, J. Dix, E. Prestat, S. J. Haigh, I. V. Grigorieva, P. Carbone, A. K. Geim and R. R. Nair, *Nat. Nanotechnol.*, 2017, **12**, 546.
27. S. Park, K.-S. Lee, G. Bozoklu, W. Cai, S. T. Nguyen and R. S. Ruoff, *ACS Nano*, 2008, **2**, 572.
28. Z. An, O. C. Compton, K. W. Putz, L. C. Brinson and S. T. Nguyen, *Adv. Mater.*, 2011, **23**, 3842.
29. C.-N. Yeh, K. Raidongia, J. Shao, Q.-H. Yang and J. Huang, *Nat. Chem.*, 2015, 7, 166–170.
30. L. Chen, G. Shi, J. Shen, B. Peng, B. Zhang, Y. Wang, F. Bian, J. Wang, D. Li, Z. Qian, G. Xu, G. Zhou, M. Wu, W. Jin, J. Li, H. Fang, arXiv:1610.06369.
31. S. J. Gao, H. Qin, P. Liu and J. Jin, *J. Mater. Chem. A*, 2015, **3**, 6649.
32. Y. Han, Y. Jiang and C. Gao, *ACS Appl. Mater. Interfaces*, 2015, 7, 8147.
33. P. Sun, Q. Chen, X. Li, H. Liu, K. Wang, M. Zhong, J. Wei, D. Wu, R. Ma, T. Sasaki and H. Zhu, *NPG Asia Mater.*, 2015, 7, e162.
34. P. Sun, R. Ma, W. Ma, J. Wu, K. Wang, T. Sasaki and H. Zhu, *NPG Asia Mater.*, 2016, **8**, e259.
35. R. Ma and T. Sasaki, *Adv. Mater.*, 2010, **22**, 5082.
36. R. Ma and T. Sasaki, *Acc. Chem. Res.*, 2015, **48**, 136.

37. D. Golberg, Y. Bando, Y. Huang, T. Terao, M. Mitome, C. Tang and C. Zhi, *ACS Nano*, 2010, **4**, 2979.
38. Y. Lin and J. W. Connell, *Nanoscale*, 2012, **4**, 6908.
39. C. Tan and H. Zhang, *Chem. Soc. Rev.*, 2015, **44**, 2713.
40. R. Lv, J. A. Robinson, R. E. Schaak, D. Sun, Y. F. Sun, T. E. Mallouk and M. Terrones, *Acc. Chem. Res.*, 2015, **48**, 56.
41. L. Sun, H. Huang and X. Peng, *Chem. Commun.*, 2013, **49**, 10718.
42. L. Sun, Y. Ying, H. Huang, Z. Song, Y. Mao, Z. Xu and X. Peng, *ACS Nano*, 2014, **8**, 6304.

CHAPTER 5

Mechanisms of Selective Mass Transport through Graphene Oxide Membranes

ZHIPING XU

Department of Engineering Mechanics, Tsinghua University, Beijing 100084, China
Email: xuzp@tsinghua.edu.cn

5.1 Introduction

5.1.1 Selective Fluidic Transport in Nanoscale Channels

Selective fluidic transport is the fundamental process for a wide range of applications from filtration and separation in environmental technology, to chemical and biomedical engineering. Selectivity and permeability are the two key factors to assess the performance of a fluidic channel. Polymeric membranes are widely used in these fields, due to their low cost, facile fabrication and functionalization procedures. The transport pathways defined by the polymeric networks are rather complex (Figure 5.1a), with the selectivity S controlled though the size-sieving mechanism and chemical interaction between the fluid and polymeric backbones. For high selectivity, one may aim to design narrow channels with size comparable with that of the species to be rejected and functionalize the channel wall for low adhesion.

Nanoscience & Nanotechnology Series No. 47
Graphene-based Membranes for Mass Transport Applications
Edited by Pengzhan Sun and Hongwei Zhu

Published by the Royal Society of Chemistry, www.rsc.org

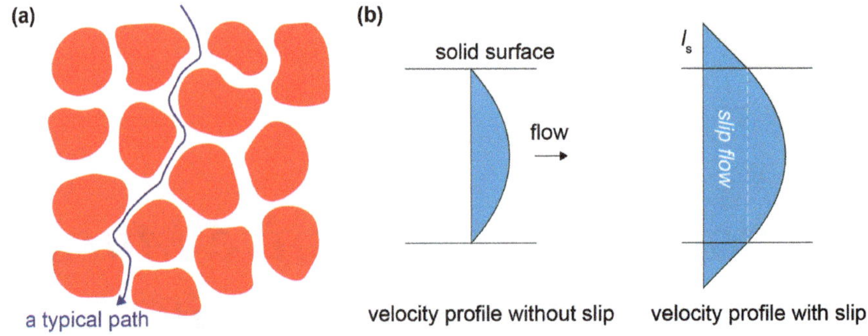

Figure 5.1 (a) The typical transport pathway in porous media. (b) Paradigm shift of fluidic transport mechanisms in macro- and nanochannels, without and with boundary slip.

The permeability P of a membrane measures the ratio between the flow rate q and pressure gradient $\mathrm{d}p/\mathrm{d}x$, which can be related to the geometrical factor f/τ. Here, the porosity f is defined as the relative volume of a membrane that can be occupied by the fluid, while the tortuosity τ is the ratio between the average length of the flow path and the membrane thickness. High-density arrays of straight channels across the membrane are thus preferred for high permeation. According to the law of viscous flow, the resistance measured by the ratio between pressure gradient and the averaged flux scales as r^{-2} for a cylindrical channel or d^{-2} for laminar flow between two parallel plates, where r is the radius of the channel and d is the distance between the plates. Consequently, wider channels are preferred for higher permeability that contradicts with the preference for higher selectivity, and thus the performance trade-off has to be well addressed in membrane design.

With the length scale of a fluidic channel shrinking down to the nanoscale, the selectivity S is elevated to the molecular level, which holds great promise in nanofiltration and (reversed) osmotic separation. The geometry of nanochannels and the nature of the fluid–channel interaction (at the wall and entrance/exit) can be fully utilized to tailor the selectivity. By reducing the width of channels, the fluid under nanoconfinement becomes more structured due to the wall adhesion, demonstrating a layered order typically, and the viscosity of fluid usually increases as a result. From a non-slip picture of pressure-driven viscous flow, the gradient of the velocity profile across the channels is parabolic, and the flow resistance is enhanced as the channel width decreases, leading to the reduction in P. However, for channels with atomistically smooth surfaces, significant boundary slip could be activated for the flow due to the ultralow (solid–liquid) interfacial friction, resulting in a much smoother velocity profile (Figure 5.1b). Moreover, the capillary pressure of nanoscale liquid condensation could provide an additional driving force to enhance P.

5.1.2 Graphene Oxides and Graphene Oxide Membranes

Recent studies have not only elucidated the underlying mechanisms of selective mass transport, but also demonstrated remarkable enhancements in *S* and *P* in specific nanochannels such as carbon nanotubes.[1] However, fabricating nanoscale channels is challenging by conventional top-down techniques such as etching and lithography, because of the lack of atomic-scale precision control, low yield and high cost. In comparison, constructing membranes from nanostructures (*e.g.* nanowires, nanotubes, nanosheets) is more promising for practical applications.

Graphene oxide (GO) is the most widely studied graphene derivative in the family of functionalized graphene. With oxygen-rich functional groups present on the basal plane and at the edges (Figure 5.2a), negatively-charged GO sheets demonstrate hydrophilicity and thus could be well dispersed in water. The atomic structures of GO strongly depend on the preparation procedure, and the GO sheet is a corrugated, one-atom-thick membrane due to the presence of functional groups and defects therein (Figure 5.2b and 2c). Ultra-high-resolution transmission electron microscopy (TEM) studies show that the pristine, oxidized regions and nanopores in a typical GO sheet have areal percentages of approximately 16%, 82%, and 2%, respectively.[2–4] Several atomic models have been proposed to understand the properties of GO, based on a periodic lattice in the early days. Disordered models with specific C/O ratios were then devised based on updated experimental evidence, with hydroxyl and epoxy groups on the basal plane, and carboxylic groups at the edges. One of the most widely used models within this spectrum is the non-stoichiometric Lerf-Klinowski structure. A more recent development after the rise of graphene is Gao's model that takes into account the five- and six-membered lactol rings, ester of a tertiary alcohol, hydroxyl, epoxy and ketone functionalizes (Figure 5.2a).[5] It should be noted that the chemistry of GO is metastable at room temperature, it degrades within a characteristic relaxation time of one month,[6] and it could further be modulated in a light or atmospheric environment.

The GO dispersion can be processed into thin films by filtration or evaporation, which are usually micrometer-thick for the consideration of uniformity and permeability.[7] Electron microscopy (EM) and X-ray diffraction (XRD) studies show that the film features a layer-by-layer order (Figure 5.3a and 3b). Considering the finite size of each GO sheet of ~0.1–1 μm, it was widely assumed that there is a brick-and-mortar hierarchy as identified in clays and biological composite structures.[8] Consequently, the transport pathway for the cross-membrane flow in a GO membrane is highly anisotropic, composed of the 2D territory confined between neighboring GO layers, the nanopores within the GO sheet, and the inter-edge slits that connect neighboring territories. While for the parallel (in-plane) flow, mass transport in the interlayer gallery controls the whole process.

GO membrane models with simplified microstructures have been used in recent studies of fluidic transport in GO membranes[10] (Figure 5.3c).

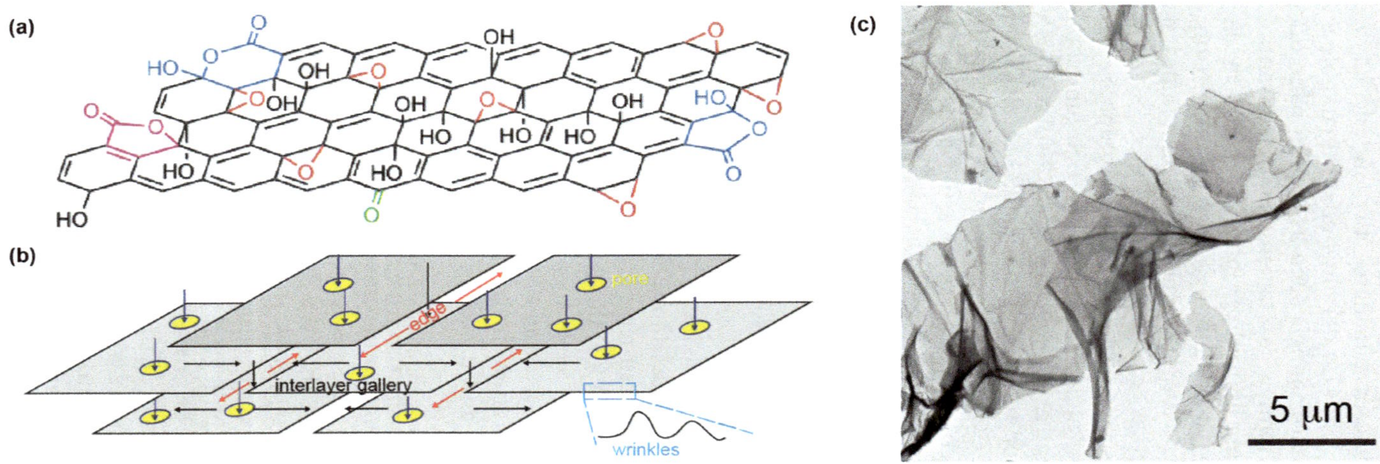

Figure 5.2 (a) Models proposed for the atomic structures of graphene oxide.[6] Reprinted by permission from Macmillan publishers Ltd: *Nature Chemistry* (W. Gao, L. B. Alemany, L. Ci and P. M. Ajayan, 2009, *Nat. Chem.*, **1**, 403–408) Copyright 2009. (b) Microstructures of graphene oxide membrane. Reproduced with permisison from N. Wei, X. Peng and Z. Xu, (2014) Understanding water permeation in graphene oxide membranes, *ACS Appl. Mater. & Interf.* **6**(8):5877–5883, Copyright 2014 American Chemical Society.[11] (c) Morphologies of graphene oxide sheets.

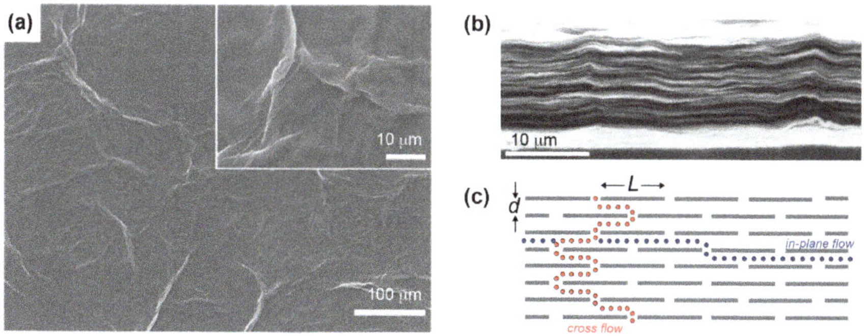

Figure 5.3 (a, b) The microstructures of GO membranes, top and side views.[7,9] Reproduced from ref. 7 and ref. 9 with permission from John Wiley and Sons, © 2015 WILEY-VCH Verlag GmbH & Co. KGaA, Weinheim. (c) The simplified 2D model for GO membranes that has been widely used to study the fluidic transport processes, which, however, oversimplifies the 3D nature of the transport pathway.

However, very little is quantitatively known about their microstructural features, due to the complexity that prohibits reliable experimental determination of the hierarchical structure across multiple length scales. However, several key microstructural characteristics have been neglected in these idealized models, which can be critical in understanding and predicting the selective fluidic processes in GO membranes. For example, the edge alignment between GO sheets and the two-dimensional stacking patterns of sheets determine the tortuosity. By assuming a uniform size distribution of the GO sheets and a random distribution of the edge alignment, and considering the membrane contains nanopores or slit-like inter-edge spacings with a width of w and an areal density of n, the probability to find a straight path across an N-layer membrane then decreases drastically as $2(wn)^{N-1}$.[11] Moreover, the GO sheets are not strictly flat within the membranes. Ripples, wrinkles and stacking disorder exist, which form short-cuts to the transport pathways for both cross-membrane and in-plane flow, and modify the deterministic factors for both selectivity and permeability.

5.2 Mechanisms of Selective Mass Transport

5.2.1 Liquid

Water. With nanoconfined channels and engineerable chemistry at the single-sheet level, the GO membranes have been widely explored in selective liquid-phase transport. Water is a polar fluid that prefers to form a hydrogen-bond network in its bulk form at ambient conditions. When a GO membrane is immersed in aqueous solution, the layer-stacked GO sheets tend to swell by adsorbing water molecules into the hydrophilic channels. The interlayer distance will increase from its initial value $d = \sim 0.6$ nm in the dry state to 6–7 nm after ~120 h. Moreover, this interlayer distance decreases with the ionic strength of solution, which is ~2 nm for 100 mM NaCl and Na_2SO_4 solutions.[12] As the GO sheet consists of both pristine and oxidized regions, the expansion in the *d*-spacing leaves open space between the pristine regions to accommodate guest molecules.

The microstructural evolution of GO membranes in aqueous solution has strong implications in the fluidic transport processes through the nanochannels therein. Experimental evidence shows that the functional groups and defects created by oxidization are not evenly distributed in the GO sheet (Figure 5.4a and 4b). The pristine regions could form a percolated network surrounding defective regions[4] or be isolated by the continuous region of functional groups or defects.[13] The size of pristine graphene

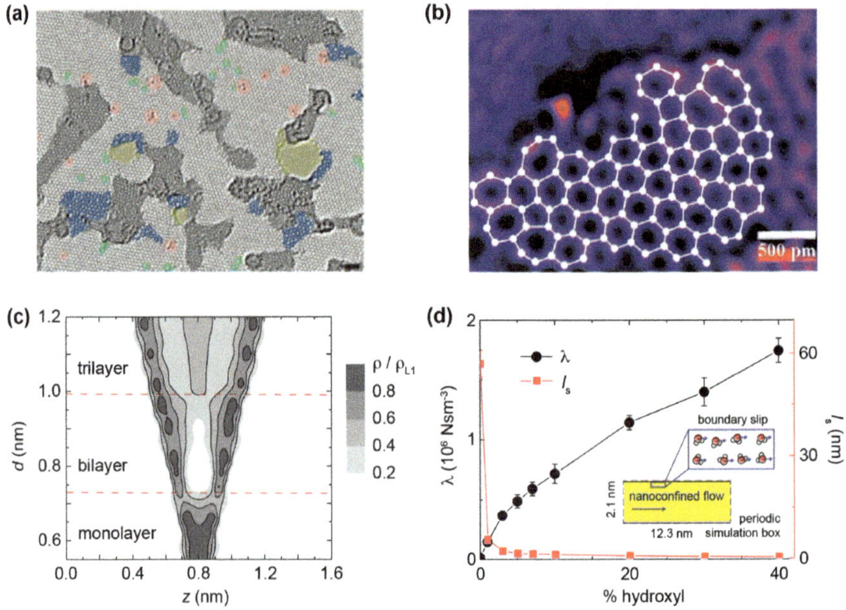

Figure 5.4 (a, b) Pristine and defective regions of GO showing pristine graphene regions percolated in the GO sheet, surrounding nanopores, oxidized and disordered regions (panel a)[4] or discrete pristine patches isolated by the continuous defective region (panel b).[13] (c) The density distribution of water confined between parallel graphene sheets with an interlayer distance *d*. (d) The friction coefficient λ and slip length l_s for water flow between graphene sheets functionalized by hydroxyl groups.[15] Panel A reprinted with permission from C. Gómez-Navarro C, J. C. Meyers, R. S. Sundaram, A. Chuvilin, S. Kurash, M. Burghard, K. Kern and U. Kaiser, (2010) Atomic structure of reduced graphene oxide. *Nano Lett.*, **10**(4), 1144–1148, Copyright 2010 American Chemical Society. Panel B reprinted with permission from S. H. Dave, C. Gong, A. W. Robertson, J. H. Warner and J. C. Grossman, (2016) Chemistry and structure of graphene oxide *via* direct imaging, *ACS Nano*, **10**(8), 7515–7522. Panels C and D reprinted with permission from N. Wei, X. Peng and Z. Xu, Physical Review E, **89**, 012113, 2014, Copyright 2014 by the American Physical Society.

regions or clusters of defects is ~1–2 nm. The van der Waals interaction between water and pristine graphene is weak, and could be significantly enhanced by the presence of oxygen-rich groups and defects that break the charge neutrality of graphene. As a result, the interfacial friction for water flow is much lower in the pristine region compared to the defective region. If a percolated transport pathway containing pristine graphene lattice only is formed in the 2D interlayer gallery, high permeability could be achieved.

Theoretical studies have shown that nanoconfined water between graphene or GO sheets displays a layered structure, at an interlayer distance below $d = \sim 2$ nm (Figure 5.4c). For monolayer water with $d = \sim 0.65$ nm, even crystalline (ice) structures are identified at ambient conditions.[14] The interfacial friction force between the absorbed solid-like water layer and graphene is extremely low, leading to significant boundary slip and modification of the velocity profile. As a result, the flow rate is enhanced compared to the prediction for non-slip Poiseuille flow between two parallel plates, that is, $q_{ns} = -d^3\Delta PW/(12\eta L)$. Here, L is the channel length, W is the width of cross section, and d is the interlayer distance or channel width that is redefined here as the hydrodynamic thickness by excluding the depletion region between water molecules and the carbon basal plane in GO, with a typical thickness of $\delta = 0.5$ nm. η is the dynamic viscosity and $\Delta P/L = dp/dx$ is the applied pressure gradient. With the additional contribution from boundary-slip flow to the water flux, the flow enhancement factor is $\varepsilon = q_s/Q_{ns} = 1 + 6l_s/d$. That is to say, with a slip length of ~1 nm, the flux in the GO channel could be doubled.

The slip length for the water interface with pristine graphene is $l_s = 48$ nm, which, however, can be significantly reduced by the presence of functional groups and defects. The presence of 5% hydroxyl or epoxy groups results in a two-order reduction in l_s, to 0.44 nm, and thus no enhancement would be expected (Figure 5.4d).[15] Consequently, from the simplified 2D model of GO membrane (Figure 5.3c), a percolated pathway with channels confined between pristine graphene regions has to be formed for the flow enhancement offered by the boundary slip. However, in GO membranes, the pristine graphene channels are surrounded by oxidized regions, which could result in a side-pinning effect that weakens the enhancement if the pristine channels are narrower than a few nanometers.[11] Moreover, if a percolated pristine graphene channel cannot be formed within the interlayer space, the water has to flow across defective regions with high flow resistance, and thus P would be reduced.

The interlayer gallery of a GO membrane contains the main transport pathway for in-plane water flow, and it makes the dominant contribution to the cross-membrane flow if the membrane is thick, considering the high aspect ratio of GO sheets, that is, $L/d \sim 10^3$. Here, L is the lateral dimension of a GO sheet and d is the interlayer distance. Nanopores within the GO sheets or slits between graphene edges form cross-plane channels or shortcuts in the transport pathway. The resistance for water flow through the nanopores or the inter-edge spaces is similar, showing no significant boundary slip and no enhanced water flow in consequence. The cross-plane flow becomes more dominant in the transport pathway for thin GO membranes, and the permeability could be elevated due to the highly reduced tortuosity, although the mechanical strength of the membrane is sacrificed by reducing its thickness. Ultrathin GO membranes demonstrate a high pure water flux of 21.8 $Lm^{-2} h^{-1} bar^{-1}$, a high retention for organic dyes >99%, and a moderate retention of ~20–60% for ion salts,[16] where both physical sieving and electrostatic interaction between the molecular species and charged groups in GO control the rejection process. As the membrane thickness increases from 22 to 53 nm, the pure water flux across the membrane decreases dramatically from 21.8 to 3.26 $Lm^{-2} h^{-1} bar^{-1}$, clearly demonstrating the increased contribution from in-plane flow in thick membranes, compared to the cross-membrane transport through nanopores and nanoslits.

As the channel width increases beyond 2 nm, the viscous dissipation in nanoconfined water flow becomes significant, and the contribution of boundary slip becomes relatively small.[17] Although the flow is less enhanced in such a regime compared to narrower channels, the overall contribution to the water flux increases under the same pressure gradient due the enlarged cross section area of channels. Consequently, the existence of nanochannels in the GO membrane, resulting from intrinsic wrinkles of the GO sheets, or created from nanowire templates, could elevate the permeability, and without quantitative determination of the membrane microstructure, it is difficult to link the selectivity and permeability of a membrane to the molecular transport processes in the nanochannels.

The entrance/exit loss and capillary driven force should also be taken into account to estimate the overall permeability of the GO membrane in practical situations, especially for the vapor-phase water transport. A recent study showed that submicrometer-thick GO membranes are completely impermeable to liquids, vapors, and gases including He but allow unimpeded permeation of water, which is at least 10^{10} times faster than He.[10] The measurements were carried out by analyzing the evaporation rates of molecular species across the membrane from their weight loss. Impressively, the unimpeded water permeation in GO membranes is as fast as water evaporation through an open aperture, suggesting an evaporation-limited regime. The fast water permeation is explained by the presence of a high capillary pressure formed between GO sheets, and a large-area, percolated path of pristine graphene regions, allowing water flow with ultralow friction. Compared to the as-prepared GO, the interlayer distance in reduced GO by thermal annealing decreases significantly from ~1 to ~0.4 nm. As a result, the water permeability is reduced remarkably by 100 times, demonstrating the significance of in-plane transport within the interlayer gallery and the critical role of interlayer distance control in the molecular sieving process.

Organics. Efficient separation membranes for organics are highly demanded in crude oil processing and chemical synthesis in the biochemical and pharmaceutical industries.[18] GO membranes are physically stable in

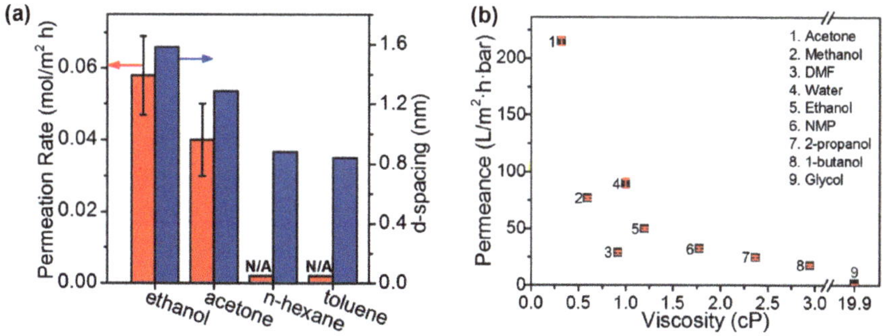

Figure 5.5 (a) The permeation rate of pyrene and *d*-spacings of solvent-soaked GO membranes in different solvents.[9] (b) The permeance of organic solvents plotted as a function of the solvent viscosity.[19]
Reproduced from ref. 9 and ref. 19 with permission from John Wiley and Sons, © 2016 WILEY-VCH Verlag GmbH & Co. KGaA, Weinheim.

various organic solvents (*e.g.* acetone, ethanol, toluene and n-hexane) with polarities weaker than that of water, and they do not swell in nonpolar solvents. The *d*-spacing of GO membranes that is critical for the molecular sieving process can thus be tuned by the choice of organic solvents. The selectivity between organic fluids in GO membranes is achieved through the difference in their permeability, where the size or steric sieving mechanism applies, as well as the contrast in their interactions with GO sheets (Figure 5.5a). Thermal annealing of GO sheets increases the areal percentage of pristine regions and thus enhances the π–π interaction between the sp^2 regions of GO and specific molecules such as pyrene. As a result, the permeation rate of pyrene decreases rapidly by increasing the annealing temperature, and the GO membrane could even be blocked after thermal annealing at 180 °C.[9]

It was also reported that the reduced GO membrane is stable in organic solvents (*e.g.* methanol, acetone, and dimethylformamide) and in strong acidic (*e.g.* sulfur acid), alkaline (*e.g.* potassium hydroxide) or oxidative (*e.g.* nitric acid) media. The permeability of organic solvents through the reduced GO membrane has a strong correlation with their viscosity, where P decreases with η in general, which is consistent with the Poiseuille prediction (Figure 5.5b).[19] The membrane shows high rejection to small molecules with charges the same as that of GO, or neutral molecules with sizes larger than 3.4 nm, while retraining its high permeability to organic liquids. Moreover, the performance can be further improved or modulated by modification with ionic polyelectrolytes, showing great potential for organic solvent nanofiltration and (reverse) osmosis applications.

5.2.2 Gas

In contrast to pressure-driven filtration, where molecules are sieved by their hydrodynamic size, gas transport through a membrane is usually a coupled sorption–diffusion process driven by the chemical potential gradient. The permeability P can be expressed as a product of the sorption and diffusion coefficients (S and D), and the selectivity is determined by the difference in P.[20] The gas transport mechanism changes from Knudsen diffusion for mesopores (following the IUPAC terminology, with pore sizes from 2 to 50 nm) to constrained molecular diffusion for micropores (<2 nm). The Knudsen model can be applied as the mean free path l_{MFP} of gas molecules is longer than the pore size. In this regime, molecular diffusion is limited by thermalized collisions with the wall, and the permeability is $2cd/3(8RT/\pi m)^{1/2}$, where c is the molar mass, d is the pore size, and m is the molecular weight. R is the ideal gas constant and T is the temperature. Selective gas transport in this regime is thus controlled by the molecular weight (m). While for constrained molecular diffusion, the effect of molecule–wall interactions becomes significant.[21]

In a GO membrane, gas diffusion is controlled by the size and shape of the molecules and nanochannels, as well as the interaction between molecular species and the GO sheets. The interlayer spacing and chemistry of the GO sheet could thus be engineered for highly selective gas permeation.[22] Rapid transport of gases in carbon nanotubes was reported, with transport diffusivities orders of magnitude higher than those in the zeolites with comparable pore sizes. This high permeability is attributed to the atomic smoothness of the graphitic walls.[23] However, in GO membranes, the functional groups, defects in the GO sheets and their corrugation reduce the gas permeability compared to that between pristine graphene sheets with the same channel width.[22]

GO membranes with thickness down to 1.8 nm demonstrate high selectivities for H_2/CO_2 (3400) and H_2/N_2 (900) separation, which are 1–2 orders higher in magnitude than that of the state-of-the-art microporous membranes with pore size <1 nm.[24] The measured permeability is closely related to the kinetic diameter of gas molecules. Hydrogen molecules with a kinetic diameter of 0.289 nm permeate ~300 times faster than CO_2 (0.33 nm) through an 18-nm-thick GO membrane at 20 °C. The slight difference between their kinetic diameters suggests the critical mesh size of the pore is ~0.289–0.33 nm. O_2 and N_2 show similar permeabilities to CO_2, while CO and CH_4 have a slightly higher permeability although their kinetic diameters are larger. Single-gas transport measurements show that the narrowed interlayer space in reduced GO membranes does not change the permeability, and there is an exponential dependence between the permeance and membrane thickness, indicating that the molecular transport pathway is mainly through structural defects in the GO layers.[24]

Few- and several layered graphene and GO membranes were fabricated, both displaying excellent gas separation performance, but the mechanism of selective transport depends on the microstructures.[25] The few-layer graphene membrane is permeable for gases, with a high O_2/N_2 selectivity, suggesting that gases diffused not only through the nanopores in graphene but also through the interlayer gallery with wrinkles and disorder in the graphene stacks. For GO membranes with non-interlocked microstructures, the gas transport is determined by the pore size and the molecular mean free path in a gas mixture, and as in the Knudsen diffusion regime, gases are separated based on their differences in molecular weight. However, for GO membranes consisting of close-packed, interlocked GO layers, the order of gas permeability is $CO_2 > H_2 > He > CH_4 > O_2 > N_2$ at 298 K, following a molecular-sieving mechanism (Figure 5.6).[25]

The GO membranes show CO_2-philic permeation behaviors in general, as the polarity of C–O bonds in the molecule allows for interaction with polar groups in GO such as carboxylic acid. This effect holds great potential in industrial CO_2 separation processes, and can be further enhanced by the presence of water, with CO_2 acting as a Lewis acid or a Lewis base that participates in hydrogen bonding.[25] The intercalation of gas molecules in GO membranes is highly affected by the affinity between the hydrophilic surface of GO and target molecules including CO_2, CH_4, H_2 and N_2. Only CO_2 can be intercalated into the GO membrane.[26] However, the swelling of interlayer spaces in water changes the intercalation phenomena significantly. All of the gases tested could then be intercalated, but CO_2 could be most solvated, surrounded by the intercalated water molecules and partially with functional groups on GO surfaces. The dynamics of intercalated water is retarded compared to bulk water because of its interaction with the GO surface; enhanced gas storage could thus be achieved within the intercalated water.[26]

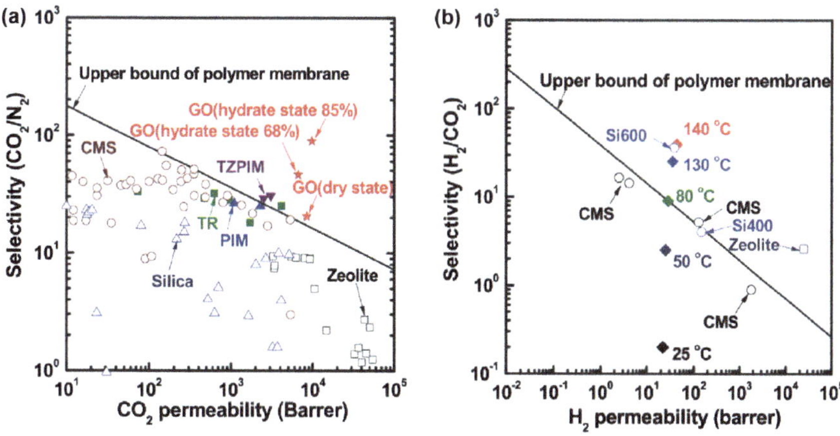

Figure 5.6 (a) Relationship between CO_2 permeability and CO_2/N_2 selectivity of GO membranes under dry and humidified conditions. (b) Comparison of gas separation performance between GO membranes and other membranes for the H_2 permeability and H_2/CO_2 selectivity.[25]
From H. W. Kim, H. W. Yoon, S.-M. Yoon, B. M. Yoo, B. K. Ahn, Y. H. Cho, H. J. Shin, H. Yang, U. Paik, S. Kwon, J.-Y Choi, H. B. Park, 2013, Selective gas transport through few-layered graphene and graphene oxide membranes, *Science* **342**(6154), 91–95. Reprinted with permission from AAAS.

The sorption mechanism in molecular transport has also been used to explain the dependence of water vapor permeability through GO membranes on the relative humidity,[10] where a linear relationship was found between the vapor permeation rate and the volume sorption, which can be explained by a quasi-equilibrium picture where the feed side is in sorption equilibrium and the sorption is determined by the relative humidity within the sealed volume. The membrane operates in the stationary regime and the rate-limiting step of permeation is the diffusion through the membrane as the opposite surface layer is facing zero humidity conditions or a vacuum.[27]

5.2.3 Ions

One of the major targets of water desalination, recovery and recycling processes is to selectively remove ions from the water. The ionic selectivity of a GO membrane is defined by the size and shape of ions, as well as their interaction with the hydrophilic, negatively charged surfaces. In the aqueous form, ions are gel-like as they possess surrounding hydration shells of water molecules. The strength of hydration depends on the type of ions. Small monovalent ions (Li^+, Na^+, F^-) feature higher hydration energies than larger ones (K^+, Cl^-, Br^-, I^-). The ions possessing higher charge densities also bind more strongly with larger water clusters. Moreover, the anions hold their hydration shells relatively more strongly than the cations due to the difference in the strength of ion–water interaction.[28]

For ions confined between nanochannels with pore sizes comparable to that of the hydrated ions, such as those traveling across the GO membranes, their solvation shell could be strongly disturbed. The nanoconfinement and reconstruction of the solvation shell make ionic transport through GO membranes more complicated than the process in the simple size-sieving picture. Firstly, the interaction between GO and the ions differs due to the hydration level of the ions, and the oxygen-rich groups could participate as part of the ionic solvation shell. At the same interlayer distance, electrostatic interactions between GO and small ions could be better screened and the permeation rate could then be higher.[29] Secondly, as the width of nanochannels becomes comparable or smaller than the Debye screening length, the interaction between the ions and pristine and oxidized regions starts to control the transport process, which includes not only the cation–π interaction between the main-group cations and the graphitic regions, but also the coordination between transition metal ions and the oxidized region.[29,30]

Ion transport through GO membranes was investigated using a nanofluidic device.[31] The massive parallel 2D channels between neighboring GO layers allow efficient in-plane ionic transport. The ionic conductance measured is as high as that of the bulk electrolyte solution at high salt concentration. However, at low salt concentrations below ~0.05 M, the conductance deviates from bulk behavior and displays a concentration-independent, surface-charge-governed regime of ionic transport. The negatively charged nature of GO also leads to a high affinity to the cations while anions are repelled, allowing ion separation. The flexibility of a GO paper based nanofluid device was also demonstrated by mechanical folding for 30 cycles.

There is also experimental evidence showing that the hydrated radius of salt ions in a molecular sieving regime determines the rate of ion permeation.[32] The ions with sizes below the critical mesh size of the nanochannel permeate at almost the same rate, whereas large ions and organic molecules exhibit no detectable permeation. These results suggest that, to avoid ionic conduction, the size of the interlayer channel should be controlled down to ~0.6 nm to allow monolayer water transport while rejecting even the smallest salts. Nanopores or inter-edge slits inside GO are expected to have the same critical size for ionic transport. However, a recent work showed that, as an extreme case of ionic transport, thermal protons are able to permeate even through a perfect graphene lattice with a surprisingly low resistivity of $10^{-3}\ \Omega\,cm^2$ above 250 °C. The transport can be further enhanced by decorating the graphene monolayer with catalytic metal nanoparticles that could lower the tunnel barrier.[33] Moreover, deuterons permeate through these crystals much more slowly than protons, resulting in a separation factor of ~10 at room temperature. This isotope effect is attributed to a difference of ~60 meV between the zero-point energies of incident protons and deuterons, which translates into the equivalent difference in the activation barriers posed by two-dimensional crystals.[34]

Selective ion permeation across GO membranes was studied by considering a wide spectrum of ion types (Figure 5.7). It was reported that sodium ions permeate much faster than heavy-metal salts such as Mn^{2+}, Cd^{2+}, and Cu^{2+}. Notably, copper sulfate and rhodamine B are blocked entirely because of their strong interactions with GO.[30] Specifically, the transition metal ions could be coordinated by the oxygen-rich groups on GO and thus be strongly impeded. Further investigation on alkali and alkaline earth cations showed relative transport rates in the order of $H^+ > Mg^{2+} > Na^+ > Ba^{2+}$, Ca^{2+} and K^+, from the highest value to the lowest ones, which can be explained by a combined effect of π–cation interactions and nanoconfined hydration.[29] The permeability increases as the lateral size of the GO sheet is reduced, where a shorter path is available for ionic conduction. Moreover, a remarkable temperature-dependence is observed from 20 to 40 °C. The ion penetrations increase significantly with temperature, indicating the role of thermal activation in the permeation. The ion transport through GO membrane is further demonstrated to be controllable under electromagnetic fields. Specifically, an electric field can either increase or decrease the ion transport depending on the flow direction, while a magnetic field can monotonically enhance the ion penetration.[35,36]

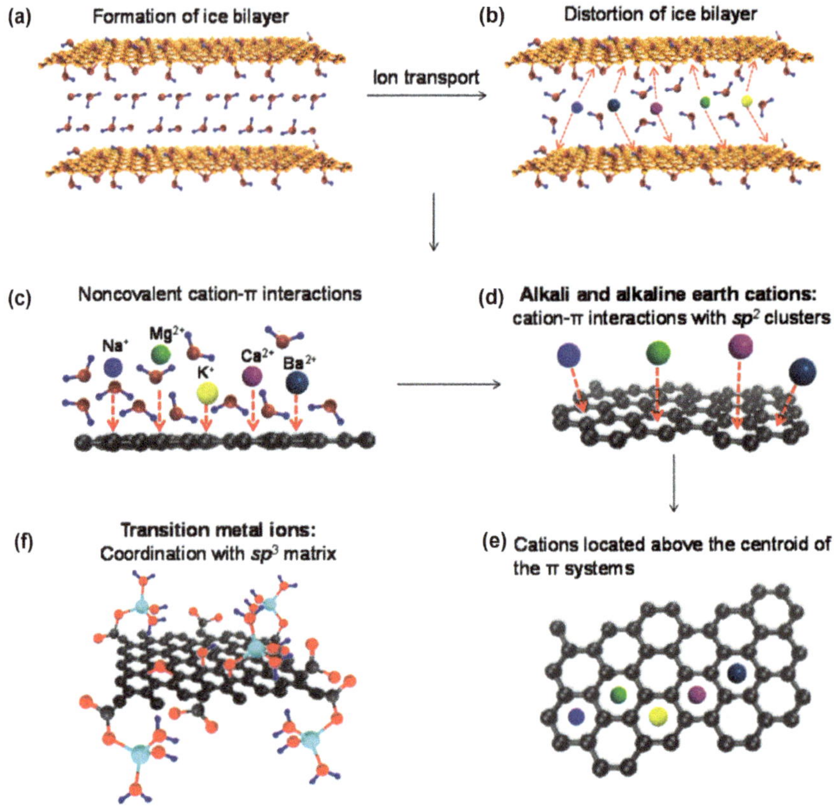

Figure 5.7 Schematic diagrams of selective ion transport through GO membranes and the diverse interactions between different cations and GO.[29] Reproduced with permission from P. Sun, F. Zheng, M. Zhu, Z. Song, K. Wang, M. Zhong, D. Wu, R. B. Little, Z. Xu and H. Zhu, 2014, Selective trans-membrane transport of alkali and alkaline earth cations through graphene oxide membranes based on cation–π interactions, *ACS Nano*, Copyright 2014 American Chemical Society.

5.3 Perspectives

The understanding of selective molecular transport mechanisms at the nanoscale is of key importance to develop high-performance filtration and separation membranes. However, the current status is far from being mature. For water transport through GO membranes, quantitative comparison between theoretical predictions and experimental measurements can hardly be made without detailed information of the microstructures. For example, the length of water transport channel can only be estimated very roughly based on the membrane thickness and average interlayer distance between neighboring GO sheets. To solve this issue, single-channel measurements should be carried out, or the fabricated membrane should have a well-defined stacking order, uniform distribution of sheet size, shape, and functionalization of the GO sheets. Computational studies provide some insights into the process, but the practical situation could be very different and complicated due to factors including the pH, salt concentration, pressure, thermal fluctuation, and electromagnetic fields, to list a few, which are difficult to consider through atomistic models and are more appropriate to be explored experimentally. In a recent work, narrow and smooth capillaries between graphene layers were fabricated through van der Waals assembly. Measurements demonstrate an unexpectedly fast water flow up to 1 $m\,s^{-1}$ driven by high capillary pressures of ~1000 bar and large slip lengths of the graphene walls. More interestingly, for channels that accommodate only a few layers of water, the flow exhibits a marked enhancement that is associated with an increased structural order in the nanoconfined water.[37] Graphene channels with a width of 3.6 nm were also found to conduct ions much faster than the SiO_2/Si nanochannels.[38] These fundamental understandings and fabrication techniques for rationally designed nanochannels constructed from graphene or GO structures lay the groundwork for the development of high-performance, multifunctional nanofluidic devices and membrane materials. Nanoporous atomically thin graphene or GO layers also present a specific type of nanochannels for selective mass (liquids, gases, ions) transport, which can be controlled by tailoring and engineering the sheets.[39]

Although impressive performance of GO membranes has been demonstrated by lab tests, there is still a gap existing towards their commercial utilization in industrial filtration and separation. To this end, large-scale fabrication of GO membranes that can maintain their structural stability in harsh environments should be established. Swelling of the GO laminates in solution should be avoided by mechanical constraints or chemical binding. The mechanical resistance of GO to applied pressure is much lower compared to the pristine graphene due to the presence of defects and surface functional groups, and it should especially be assured for ultrathin membranes that are preferred for high permeability. GO membranes could also be used as coating or barrier materials in related applications, where the supporting porous materials could provide the required stability. However, the support layer could significantly weaken the performance of GO membranes for reasons such as the internal concentration polarization (ICP), which inevitably occurs inside the support layer and leads to a dramatic decrease in membrane flux and rejection rate. As a result, free-suspended GO membranes with desired mechanical resistance are highly demanded for industrial applications. Moreover, the oxygen-rich groups could absorb either polar molecules or charged ions, resulting in a significant number of deposited residuals, which could break down the membrane performance and should be removed for regeneration.

There are promising opportunities along the direction of membrane development using monolayer or few-layer materials, which can be considered as the ultimate material design due to the fact that all the atoms in these materials are exposed to the environment for engineering and the atomic structure could be tailored by a broad class of techniques. The wide spectrum of 2D materials could endow the membrane with unconventional control and novel functions, although more mechanistic studies are needed to elucidate the link between their optoelectronic properties and nanofluidic behaviors. For example, the molecular transport process through open channels in these materials could be coupled with other processes such as electricity generation and energy harvesting. Understanding these functional processes could also inspire studies on transport and energetic processes in biological systems.

References

1. G. E. Karniadakis, A. Beskok and N. A luru, *Microflows and Nanoflows: Fundamentals and Simulation*, Springer Science & Business Media, 2006.
2. K. Erickson, *et al.*, Determination of the local chemical structure of graphene oxide and reduced graphene oxide, *Adv. Mater.*, 2010, **22**(40), 4467–4472.
3. D. Pacilé, *et al.*, Electronic properties and atomic structure of graphene oxide membranes, *Carbon*, 2011, **49**(3), 966–972.
4. C. Gómez-Navarro, *et al.*, Atomic structure of reduced graphene oxide, *Nano Lett.*, 2010, **10**(4), 1144–1148.
5. W. Gao, *Graphene Oxide: Reduction Recipes, Spectroscopy, and Applications*, Springer, 2015.
6. S. Kim, *et al.*, Room-temperature metastability of multilayer graphene oxide films, *Nat. Mater.*, 2012, **11**(6), 544–549.
7. M. Zhang, *et al.*, Multifunctional pristine chemically modified graphene films as strong as stainless steel, *Adv. Mater.*, 2015, **27**(42), 6708–6713.
8. P. Fratzl and R. Weinkamer, Nature's hierarchical materials, *Prog. Mater. Sci.*, 2007, **52**(8), 1263–1334.
9. L. Huang, Y. Li, Q. Zhou, W. Yuan and G. Shi, Graphene oxide membranes with tunable semipermeability in organic solvents, *Adv. Mater.*, 2015, **27**(25), 3797–3802.
10. R. R. Nair, H. A. Wu, P. N. Jayaram, I. V. Grigorieva and A. K. Geim, Unimpeded permeation of water through helium-leak-tight graphene-based membranes, *Science*, 2012, **335**(6067), 442–444.
11. N. Wei, X. Peng and Z. Xu, Understanding water permeation in graphene oxide membranes, *ACS Appl. Mater. Interfaces*, 2014, **6**(8), 5877–5883.
12. S. Zheng, Q. Tu, J. J. Urban, S. Li and B. Mi, Swelling of graphene oxide membranes in aqueous solution: Characterization of interlayer spacing and insight into water transport mechanisms, *ACS Nano*, 2017, **11**(6), 6440–6450.
13. S. H. Dave, C. Gong, A. W. Robertson, J. H. Warner and J. C. Grossman, Chemistry and structure of graphene oxide via direct imaging, *ACS Nano*, 2016, **10**(8), 7515–7522.
14. S. Jiao, C. Duan and Z. Xu, Structures and thermodynamics of water encapsulated by graphene, *Sci. Rep.*, 2017, **7**(1), 2646.

15. N. Wei, X. Peng and Z. Xu, Breakdown of fast water transport in graphene oxides, *Phys. Rev. E*, 2014, **89**(1), 012113.
16. Y. Han, Z. Xu and C. Gao, Ultrathin graphene nanofiltration membrane for water purification, *Adv. Funct. Mater.*, 2013, **23**(29), 3693–3700.
17. H. Huang, *et al.*, Salt concentration, ph and pressure controlled separation of small molecules through lamellar graphene oxide membranes, *Chem. Commun.*, 2013, **49**(53), 5963–5965.
18. R. P. Lively and D. S. Sholl, From water to organics in membrane separations, *Nat. Mater.*, 2017, **16**(3), 276–279.
19. L. Huang, *et al.*, Reduced graphene oxide membranes for ultrafast organic solvent nanofiltration, *Adv. Mater.*, 2016, **28**(39), 8669–8674.
20. D. R. Paul and Y. P. Yampol'skii, *Polymeric Gas Separation Membranes*, CRC press, 1993.
21. H. Verweij, M. C. Schillo and J. Li, Fast mass transport through carbon nanotube membranes, *Small*, 2007, **3**(12), 1996–2004.
22. S. Jiao and Z. Xu, Selective gas diffusion in graphene oxides membranes: A molecular dynamics simulations study, *ACS Appl. Mater. Interfaces*, 2015, 7(17), 9052–9059.
23. A. I. Skoulidas, D. M. Ackerman, J. K. Johnson and D. S. Sholl, Rapid transport of gases in carbon nanotubes, *Phys. Rev. Lett.*, 2002, **89**(18), 185901.
24. H. Li, *et al.*, Ultrathin, molecular-sieving graphene oxide membranes for selective hydrogen separation, *Science*, 2013, **342**(6154), 95–98.
25. H. W. Kim, *et al.*, Selective gas transport through few-layered graphene and graphene oxide membranes, *Science*, 2013, **342**(6154), 91–95.
26. D. Kim, *et al.*, Intercalation of gas molecules in graphene oxide interlayer: The role of water, *J. Phys. Chem. C*, 2014, **118**(20), 11142–11148.
27. M. V. Korobov, *et al.*, Sorption of polar organic solvents and water by graphite oxide: Thermodynamic approach, *Carbon*, 2016, **102**, 297–303.
28. B. Tansel, *et al.*, Significance of hydrated radius and hydration shells on ionic permeability during nanofiltration in dead end and cross flow modes, *Sep. Purif. Technol.*, 2006, **51**(1), 40–47.
29. P. Sun, *et al.*, Selective trans-membrane transport of alkali and alkaline earth cations through graphene oxide membranes based on cation-π interactions, *ACS Nano*, 2014, **8**(1), 850–859.
30. P. Sun, *et al.*, Selective ion penetration of graphene oxide membranes, *ACS Nano*, 2013, 7(1), 428–437.
31. K. Raidongia and J. Huang, Nanofluidic ion transport through reconstructed layered materials, *J. Am. Chem. Soc.*, 2012, **134**(40), 16528–16531.
32. R. Joshi, *et al.*, Precise and ultrafast molecular sieving through graphene oxide membranes, *Science*, 2014, **343**(6172), 752–754.
33. S. Hu, *et al.*, Proton transport through one-atom-thick crystals, *Nature*, 2014, **516**(7530), 227–230.
34. M. Lozada-Hidalgo, *et al.*, Sieving hydrogen isotopes through two-dimensional crystals, *Science*, 2016, **351**(6268), 68–70.

35. P. Sun, *et al.*, Electro-and magneto-modulated ion transport through graphene oxide membranes, *Sci. Rep.*, 2014, 4.
36. P. Sun, *et al.*, Intrinsic high water/ion selectivity of graphene oxide lamellar membranes in concentration gradient-driven diffusion, *Chem. Sci.*, 2016, 7(12), 6988–6994.
37. B. Radha, *et al.*, Molecular transport through capillaries made with atomic-scale precision, *Nature*, 2016, **538**(7624), 222–225.
38. W. Jung, *et al.*, A novel fabrication of 3.6 nm high graphene nanochannels for ultrafast ion transport, *Adv. Mater.*, 2017, **29**, 17.
39. L. Wang, *et al.*, Fundamental transport mechanisms, fabrication and potential applications of nanoporous atomically thin membranes, *Nat. Nanotechnol.*, 2017, **12**(6), 509–522.

CHAPTER 6

Mass Transport Properties of Composite Membranes Containing Graphene Oxide Nanosheets

YULONG YING AND XINSHENG PENG*

State Key Laboratory of Silicon Materials, Department of Materials Science and Engineering, Zhejiang University, Hangzhou 310027, P. R. China
*Email: pengxinsheng@zju.du.cn

6.1 Introduction

For an ideal separation membrane, it should be thin and have robust mechanical strength to maximize permeation, have a uniform pore size distribution to retain its high selectivity, be chemically inert, and resist fouling for a long service life.[1–3]

As the famous member of the two-dimensional (2D) materials that were honoured with a Nobel Prize,[4] graphene was born for separation due to its monolayer structure, being the thinnest membrane so far with only one carbon atom thickness.[5] However, defect-free graphene was found to be an impermeable material even for the smallest molecule, helium (Figure 6.1a–e).[6] Hence, it is necessary to either drill nano-size pores in graphene nanosheets or assemble the nanosheets into a laminar membrane to endow graphene-based membrane separation properties (Figure 6.1f–i).[3,7]

Nanoscience & Nanotechnology Series No. 47
Graphene-based Membranes for Mass Transport Applications
Edited by Pengzhan Sun and Hongwei Zhu

Published by the Royal Society of Chemistry, www.rsc.org

Neither of these challenges has been easy to overcome for large-scale production until now.

As a derivative, graphene oxide (GO) was developed towards practical application in the separation field, owing to its unique structures and components, a laminar structure surrounded with a variety of oxygen-containing functional groups.[1,8–24] Due to the existence of abundant oxygen-containing functional groups within GO nanosheets, it is facile to assemble them into laminar structure membranes by spray coating, vacuum filtration, dip coating, spin coating or layer by layer self-assembly

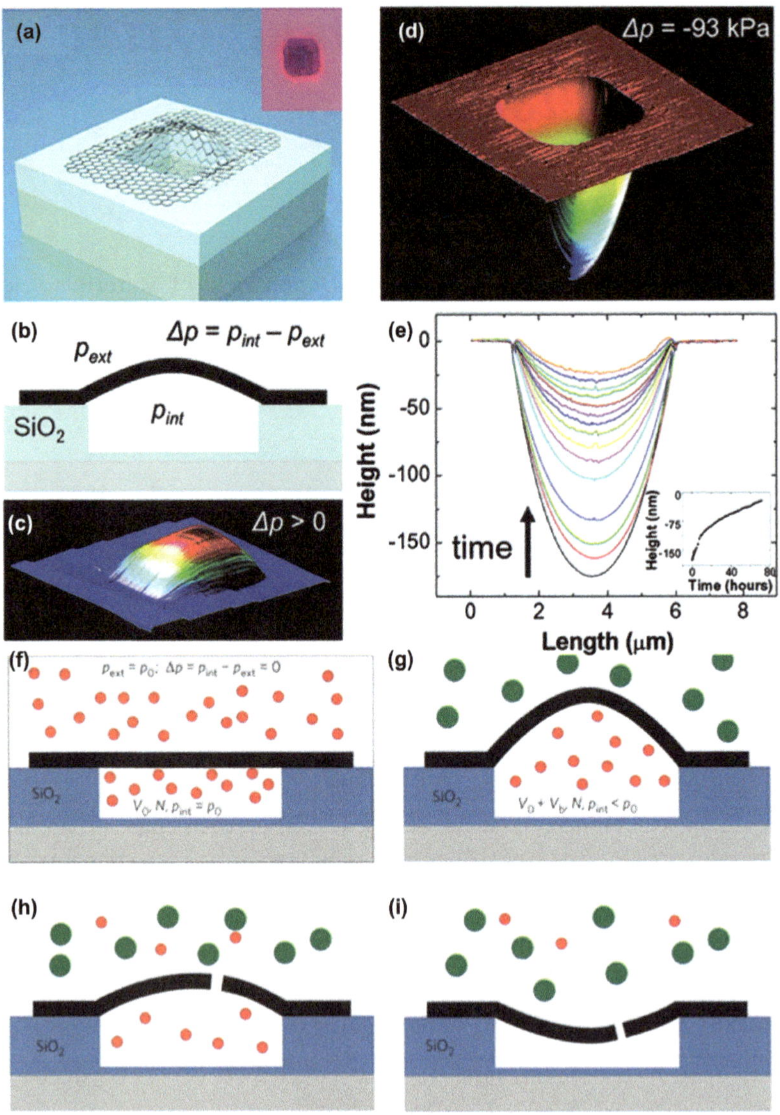

approaches.[3,25] On the other side, except for this advantage, the oxygen-containing functional groups also result in low structural and physico-chemical stability when immersing in aqueous solution. The integrity of the pure GO membrane is solely maintained by hydrogen bond interactions between the oxygen-containing groups of GO nanosheets.[26] When immersing in aqueous solution, the GO membrane readily disintegrates in water owing to electrostatic repulsion and extremely hydrophilic interactions.[18,27] This membrane assembly process may also cause problems in the precise control of membrane structure and repeatability of separation performance. And, the existence of inevitable non-selective defects within the membrane may sometimes seriously affect the separation performance, especially for this ultrathin membrane. What's more, freestanding GO membranes without any substrates or brittle GO membranes with no reliable modules are difficult to use in practical applications.

Usually, the typical interlayer spacing (d) in GO membranes formed by vacuum filtration is about 10 Å in the dry state prior to X-ray diffraction (XRD) studies, which can be tuneable to a higher value of 14.2 Å by hydration.[28,29] Geim *et al.* carried out the pioneer study of GO membrane in ion separation. This indicated that smaller ions permeate through the membranes at rates thousands of times faster than what is expected for simple diffusion while all solutes with hydrated radii larger than 4.5 Å are blocked owing to their high selectivity based on the size sieving effect.[30] Then, Sun *et al.* further illustrated that the strong coordination between heavy-metal ions with oxygen functional groups and the different strengths of cation–π interactions are also responsible for the final selective ion permeation properties (Figure 6.2).[31,32] The effects of oxidation level and flake size of the GO nanosheets on the transport characteristics were also investigated by Paneri *et al.* with the results that a smaller size of GO flake and a higher oxidation level benefit the permeability.[33] With the large amount of experiments and atomistic simulations, all the results demonstrated that the

Figure 6.1 (a) Schematic of a graphene-sealed microchamber. (Inset) Optical image of a single atomic layer graphene drumhead on 440 nm of SiO_2. The dimensions of the microchamber are 4.75 μm×4.75 μm×380 nm. (b) Side-view schematic of the graphene-sealed microchamber. (c) Tapping mode atomic force microscopy (AFM) image of an ~9 nm thick many layer graphene drumhead with $\Delta p > 0$. (d) AFM image of the graphene sealed microchamber of Figure 1a with $\Delta p = -93$ kPa across it. The minimum dip in the z direction is 175 nm. Reprinted with permission from J. S. Bunch, S. S. Verbridge, J. S. Alden, A. M. van der Zande, J. M. Parpia, H. G. Craighead and P. L. McEuen, *Nano Lett.*, 2008, **8**, 2458, Copyright 2008 American Chemical Society.[6] Schematic of a microscopic graphene membrane on a silicon oxide substrate (f) under equilibrium and (g) pressure difference. Schematic of a microscopic graphene membrane with a pore on a silicon oxide substrate (h) under positive pressure and (g) negative pressure with separation performance. Reprinted by permission from Macmillan Publishers Ltd: *Nature Nanotechnology* (ref. 7) Copyright 2012.

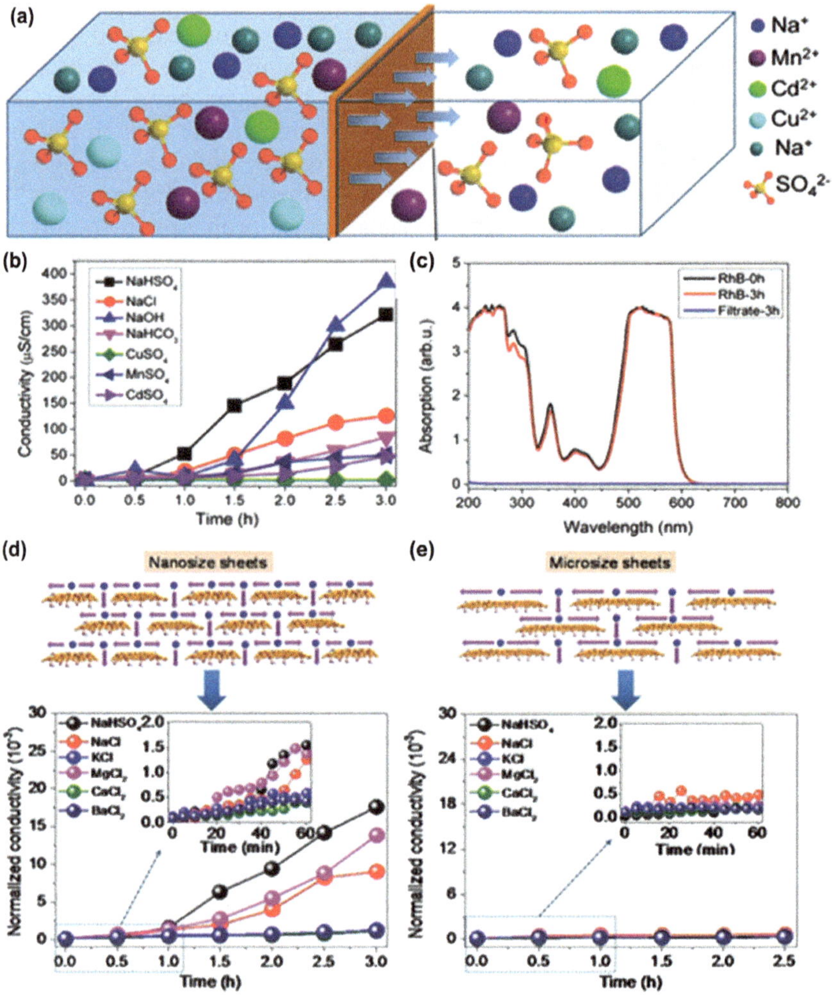

Figure 6.2 Selective mass transport through freestanding GO membrane. (a) Schematic diagram for ion permeation through GO membrane. (b) Permeations of various salt solutions (0.1 M). (c) UV–vis absorption spectra of a RhB solution (0.1 mg mL^{-1}) and its corresponding source and drain solutions after 3-h transmembrane permeation. Reprinted with permission from P. Z. Sun, M. Zhu, K. L. Wang, M. L. Zhong, J. Q. Wei, D. H. Wu, Z. P. Xu and H. W. Zhu, *ACS Nano*, 2013, **7**, 428, Copyright 2013 American Chemical Society.[31] (d) and (e) Schematic drawings of the GO laminates composed of nanometer- and micrometer-sized GO flakes, respectively, and the corresponding transmembrane permeations of alkali and alkaline-earth cations. Reprinted with permission from P. Z. Sun, F. Zheng, M. Zhu, Z. G. Song, K. L. Wang, M. L. Zhong, D. H. Wu, R. B. Little, Z. P. Xu and H. W. Zhu, *ACS Nano*, 2014, **8**, 850, Copyright 2014 American Chemical Society.[32]

highly selective separation performance of GO membranes can be attributed to the abundant nanochannels within the as-assembled membranes, including the interlayer gallery, expanded channels such as wrinkles of inter-edge spaces, defect-induced nanopores in nanosheets and the functional groups on the surface.[9,24,30–32,34,35]

However, the intrinsic mechanism of the unique ultrafast transport behaviour is still under further study and is yet to reach agreement with a unanimous conclusion. According to current understanding, pristine (non-oxidized) and functionalized (oxidized) regions both exist on graphene nanosheets, and water molecules transfer superfast through the non-oxidized zone within GO membranes, which is similar to the low-friction flow of monolayer water within small-diameter carbon nanotubes (CNTs) and hydrophobic nanopores.[30,35] At the same time, the oxidized regions serve as spacers for the 2D capillary networks that keep graphene planes apart, and strongly interact with intercalating water in a hydrated state, presenting no contribution to water permeation.[36] With the aid of isotope labelling, the superfast transport behaviour was further confirmed again by Sun *et al.* (Figure 6.3).[37] They indicated that liquid water can also afford an ultrafast permeation through GO membranes with diffusion coefficients 4–5 orders of magnitude greater than that in the bulk case, which is similar to the case of CNTs and water vapour in GO membranes. Recently, Geim *et al.* reported their current breakthrough, the discovery of low-dimensional ice locked between two graphene sheets at room temperature by high-resolution transmission electron microscopy (HRTEM) imaging,[38] which has been proposed to explain the fast water permeation through hydrophobic nanocapillaries, including CNTs and graphene-based membranes, although not all people agreed with their opinion and evidence.[39]

The diverse methods for nanosheet surface decoration and structure modulation, as well as the versatile strategies for membrane channel modifications, can all make the as-prepared membrane present desired separation performance.[40–44] Typically, GO membrane, with 3–5 nm nanochannels, shows a superior separation performance for small molecules, with ultrafast water permeance (50–70 liter per hour per square meters per bar, $LHM \cdot bar^{-1}$) and high molecule rejection (85–95%) (Figure 6.4).[20]

Since the first report of the pioneering work that found that submicrometer-thick laminated graphene oxide membranes can be completely impermeable to gases, including helium, but allow unimpeded permeation of water vapours,[36] the self-assembled GO membranes and their composite membranes have experienced the most intensive investigations in water purification, solvent-dehydration and desalination.[17,19,20,40–42,45–49] Meanwhile, desired gas separation characteristics were also achieved in GO membranes with precisely controlled laminar structures. As the first two pioneer works, GO membranes with distinct separation properties were reported by Science in 2013 at the same time.[11,50] It seems that different fabrication approaches inducing microstructures, together with the experiment environment, significantly affect the gas transport behaviour within

GO membrane. Kim *et al.* reported that high CO_2/N_2 selectivity through few-layered GO membranes can be achieved by controlling gas flow pathways and pores *via* different stacking methods in high relative humidity.[50] Li *et al.* described how structural defects within nanosheets endow GO membranes with separation selectivity and high H_2 permeation over CO_2 and N_2.[11]

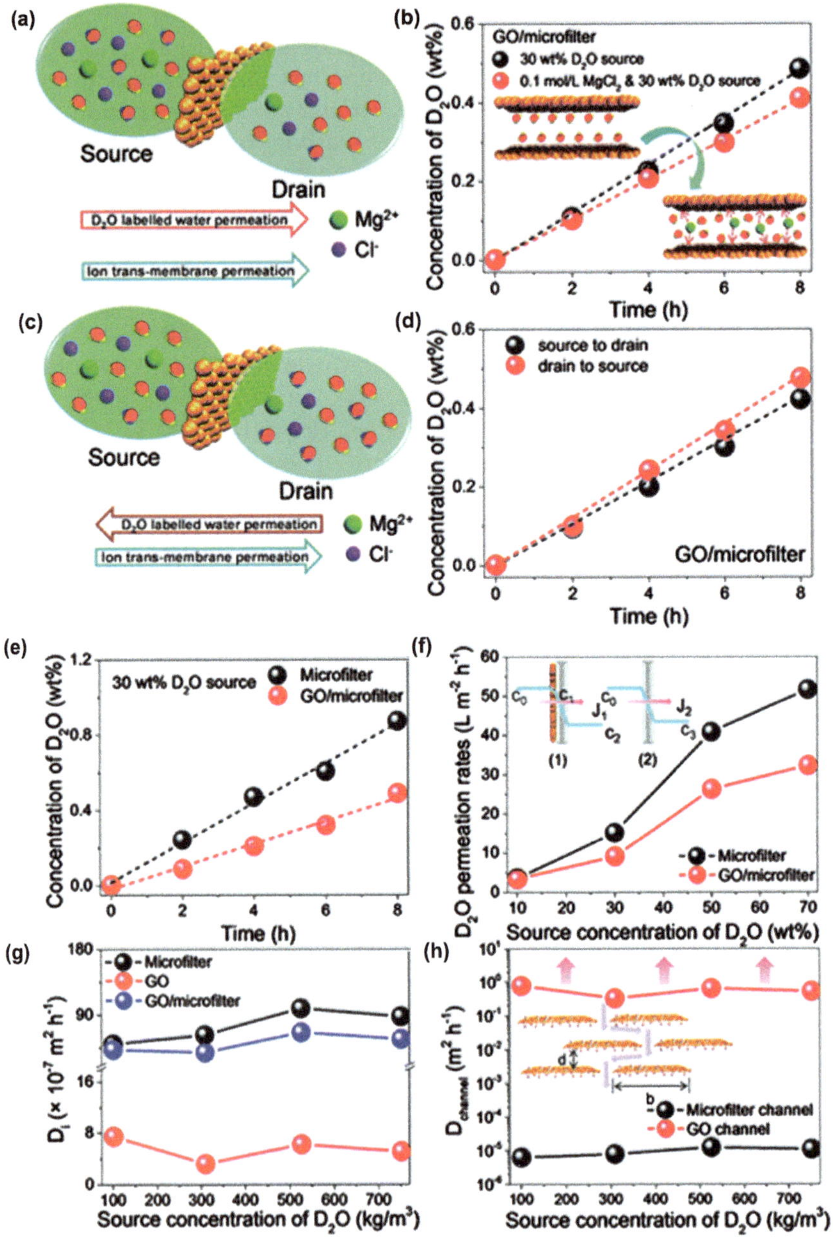

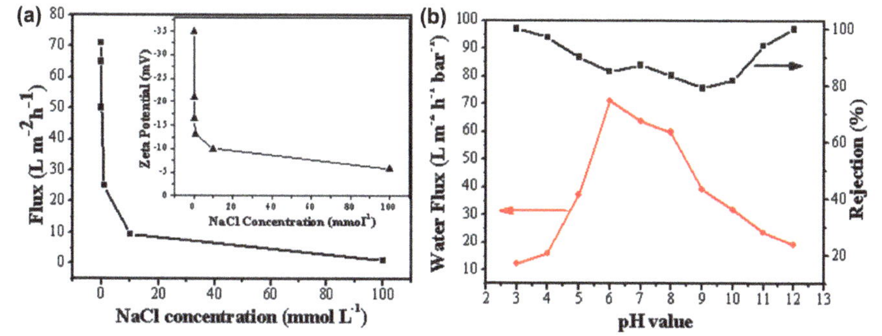

Figure 6.4 (a) The flux of NaCl solution with different concentrations; (b) the water fluxes and rejection of EB molecules at different pH through the LGO membrane. The inset in (a) is the zeta potential of GO dispersion *versus* the NaCl concentration.
Reproduced from ref. 20 with permission from the Royal Society of Chemistry.

In the following sections, we will describe in detail the advantages of GO composite membranes, such as GO nanosheets together with typical materials or nanomaterials, which expand the scope of GO membranes with excellent performance.

6.2 Mass Transport Through Graphene Oxide/ Polymer Composite Membranes

Polymeric membranes are the industry market leader both in gas and liquid separation owing to their competitiveness both in performance and

Figure 6.3 (a, c) Schematic drawings for the labelling of 0.1 M $MgCl_2$ source solution and the corresponding drain solution by 30 *wt*% D_2O tracers. (b) Water permeations through GOCM with and without ions in the sources. The inset shows the possible mechanism for water permeation in the presence of ions. (d) Water permeations through GOCM in both directions when dissolving ions in sources. (e) D_2O mass transportation *versus* time through GOCM and blank microfilter membranes with a source D_2O concentration of 30 *wt*%. (f) D_2O permeation rates through GOCM and blank microfilter membranes as a function of the source concentration. The inset shows the schematic diagrams for the calculation of diffusion coefficients through GO and microfilters by Fick's first law. (g) D_2O diffusion coefficients through the entire GO, microfilter and GOCM membranes as a function of the D_2O source concentration. (h) D_2O diffusion coefficients through the channels within GO and microfilters as a function of the D_2O source concentration. The inset shows a schematic diagram of the cross-section of GO membranes used for calculation. The red arrows indicate the lower bounds for the water diffusion coefficients.
Reproduced from ref. 37 with permission from Royal Society of Chemistry.

economics.[3] Polymeric membranes present an excellent separation performance of fast flux and high perm-selectivity, while they often show limited tolerance to extreme environments, such as high temperatures, organic solvents and corrosive environments.[51–54] Besides, the growing and special demands for membranes with larger flux, higher selectivity and flexibility never stop.

Recently, two distinct methodologies have proven effective in separation performance enhancement, combining the advantages of both polymeric membranes and GO membranes: one is to embed GO nanosheets into a polymeric matrix, and another is to utilize a polymer to modify GO nanosheets and GO membranes.

6.2.1 GO Laminate Embedded in Polymeric Membrane

6.2.1.1 Gas Separation

There has been a long history of adding graphene derivatives into a polymeric matrix for the preparation of different membranes.[45,55] Now, combination of the high gas selectivity of GO membrane and the mature manufacturing techniques of polymeric membranes has enabled the as-prepared composite membranes to break the permeability/selectivity trade-off relation in polymeric membranes.[56–60]

Jin *et al.* reported the pioneer work that involves GO nanosheet assembly in a polymeric environment for practical gas separation.[57] Once again, the interlayer spaces within GO membrane have been proven to be effective in offering molecular-sieving channels. As estimated using Geim and co-worker's method, the embedded GO laminates in random directions within a polymeric membrane largely shorten the effective transport length compared with the parallel arrangement within GO membrane, leading to faster transport behaviour. In general, two crucial properties, solubility and diffusivity, determine the molecular permeation through gas-separation membranes. Besides straight diffusion pathways, strong specific affinity and molecular sieving effects also contribute to larger flux and higher selectivity (CO_2 permeability: 100 Barrer [10^{-10} cm^3 (STP) cm/(cm^2 s cm Hg)], CO_2/N_2 selectivity: 91). Consistent with Kim's results, the as-prepared GO–PEBA membranes also allow the fastest transport for CO_2 molecules and very low permeation of other gases, with a gas permeability order of $CO_2 > H_2 > CH_4 > N_2$. In this case, the introduction of GO laminates provides a nanochannel with a size of 0.35 nm, which allows a cut-off in permeation for molecules with larger kinetic diameters, and higher CO_2 adsorption, which further accelerates the preferential CO_2 transport. Moreover, by optimizing the amount of GO nanosheets, both the CO_2 permeability and selectivity of the membranes are significantly enhanced.

With a similar method, amino acid-functionalized graphene oxide was also incorporated into a sulfonated poly(ether ether ketone) (SPEEK) polymer matrix for enhancing the CO_2 separation performance (Figure 6.5).[58]

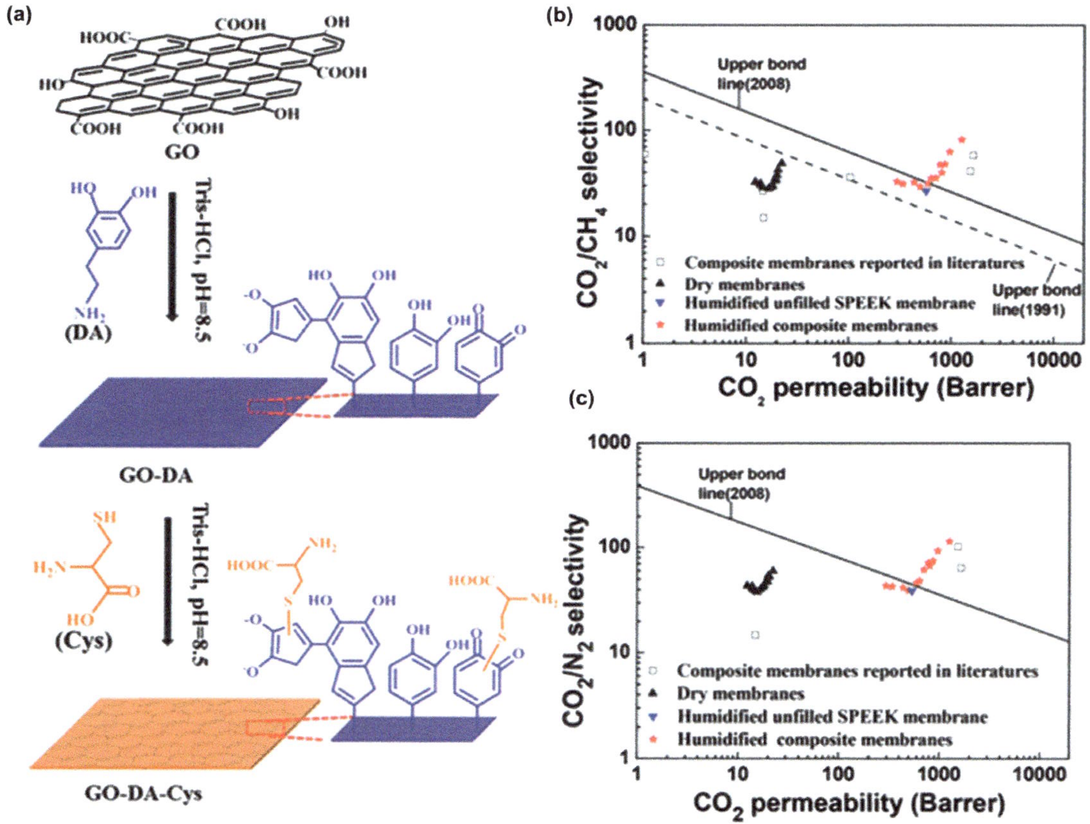

Figure 6.5 (a) The grafting mechanism of dopamine and cysteine on GO nanosheets. Pure gas separation performance of the membranes for (b) CO_2–CH_4 and (c) CO_2–N_2 mixtures, respectively (temperature: 25 °C; pressure: 1 bar). Reproduced from ref. 58 with permission from the Royal Society of Chemistry.

Amino acids with carboxylic acid and primary amine groups simultaneously enhance the solubility selectivity and reactivity selectivity, resulting in faster CO_2 molecule transport. With optimization, the composite membranes can present an excellent preferential CO_2 permeation performance (CO_2/CH_4 selectivity: 82, CO_2/N_2 selectivity: 115, CO_2 permeability: 1247 Barrer). The introduced GO nanosheets increase more tortuous paths for larger molecules, enhancing the diffuse selectivity. Besides, they also improve the mechanical and thermal stabilities of the composite membrane. Li *et al.* reported the success of a novel multi-permselective mixed matrix membrane (MP-MMM), fabricated by incorporating polyethylene glycol- and polyethylenimine-functionalized graphene oxide nanosheets (PEG–PEI–GO) into a commercial low-cost Pebax matrix, for efficient CO_2 capture (Figure 6.6).[59] The optimized MP-MMM displays excellent gas separation performance (CO_2 permeability: 1330 Barrer, CO_2/CH_4 selectivity: 45, CO_2/N_2 selectivity of 120). Wang *et al.* conducted a deeper study into the relationship between composite membrane performance and morphology of graphene oxide. GO nanosheets with different morphologies were embedded into polyurethane membranes by *in situ* polymerization, and the composites exhibited distinct CO_2/N_2 separation performances.[60] The separation performance difference is attributed to both the increase in oxygenated functional groups on the GO surface and the decrease in the number of GO layers, leading to a better distribution of GO in the polymer matrix. However, excess oxidant destroyed the structure of the GO nanosheets and formed structural defects, which depressed the separation performance of the membranes.

In conclusion, the roles of GO nanosheets within a polymeric matrix in enhancing gas separation are as follows: (a) nanochannels within GO laminates increase the length of the tortuous path of gas diffusion and generate a rigidified interface between the polymer matrix and filler, enhancing the diffusivity selectivity; (b) abundant oxygen-containing functional groups or other functionalized groups have excellent affinity for CO_2 or react reversibly with CO_2 to enhance reactivity selectivity; (c) the GO nanosheets further improve the mechanical and thermal stabilities of the composite membrane.

6.2.1.2 *Fuel Cell Exchange Membrane*

Due to its high ionic conductivity and good thermal, chemical, and mechanical stability, Nafion, a perfluorosulfonic polymer, has been widely used as a representative polymer electrolyte membrane (PEM) for fuel cells (FCs).[61] However, it has suffered from a severe methanol crossover problem occurring through swelled ionic channels.[62] Different from Nafion membranes incorporating various organic and inorganic nanofillers, which demonstrated a dramatic reduction in proton conductivity,[63] Nafion composites with GO nanosheets demonstrated much better performance, not only in the

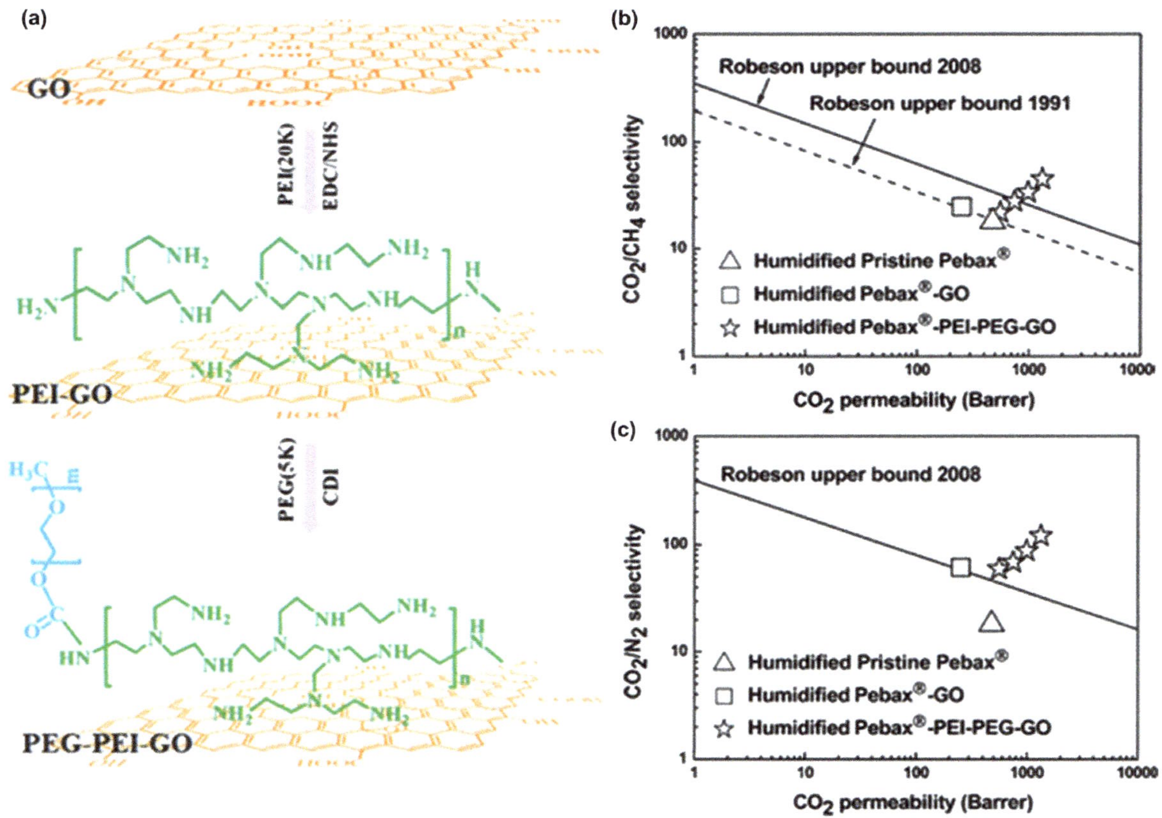

Figure 6.6 (a) Illustration of the Preparation of PEG–PEI–GO; Robeson plots for (b) CO_2/CH_4 separation and (b) CO_2/N_2 separation. Reprinted with permission from X. Q. Li, Y. D. Cheng, H. Y. Zhang, S. F. Wang, Z. Y. Jiang, R. L. Guo and H. Wu, *ACS Appl. Mater. Inter.*, 2015, 7, 5528, Copyright 2015, American Chemical Society.[59]

reduction of methanol permeability but also in the enhancement of ionic conductivity.[64]

Choi *et al.* reported the synthesis of ionic conductive Nafion/graphene oxide (GO) composite membranes for application in direct methanol fuel cells.[64] The mutual interactions of Nafion and the GO nanosheets led to the rearranged microstructures of the two domains, the hydrophobic backbone and ionic clusters. The incorporation of GO nanosheets into the ionic clusters of Nafion resulted in selective transport, decreased methanol crossover (40% less than in Nafion) while maintaining the ionic conductivity (98.1% of that in Nafion). With optimal GO loading, the Nafion/graphene oxide (GO) composite membrane demonstrated a higher power density than those of the Nafion membrane-based DMFCs. Enotiadis *et al.* also presented another functionalized GO-based Nafion nanocomposite membrane with enhanced proton transport and water retention.[65] Four aliphatic amine derivatives containing various functional groups, such as sulfonic, carboxy and hydroxy groups, were covalently bonded on the GO surfaces. Then, organo-modified GO nanofiller was incorporated into Nafion by solution intercalation. Functionalization of GO with 3-amino-1-propanesulfonic acid and 5-aminovaleric acid molecules resulted in composite membranes with high water diffusion coefficients over a wide range of temperatures (25–140 °C) and also promoted a higher hydration level with respect to filler-free Nafion. In addition to Nafion, sulfonated poly ether ether ketone (SPEEK) membrane was also decorated with sulfonated graphene oxide for direct methanol fuel cells.[66] The incorporated sulfonated graphene oxide not only increased the proton conductivity *via* increasing the number of sulfonic groups (SO_3H), but also positively affected the mechanical properties and the blocking of methanol and water molecules passing through the membrane.

The roles of GO nanosheets within polymer electrolyte membranes for fuel cell exchange membranes are as follows: (a) modulate ionic channels of polymer electrolyte membranes and decrease methanol crossover; (b) preserve or increase the proton conductivity by increasing the number of functional groups, such as sulfonic groups or amine groups; (c) improve the thermal and mechanical properties.

6.2.1.3 Liquid Separation (Ultrafiltration, Nanofiltration and Pervaporation)

The surface of membranes and their porous structure play important roles in membrane separation processes. An appropriate porous membrane for water purification and dehydration must have high permeability, good hydrophilicity and excellent chemical resistance to the feeds. However, most of the polymeric membranes are hydrophobic in nature, resulting in low permeability and fouling during aqueous separation.

Due to their unique amphiphilicity and intrinsic nanochannel structure, lots of groups have reported the addition of GO nanosheets to polymeric separation membranes, to increase membrane surface hydrophilicity, water permeability, or fouling resistance.[48,67–70] Furthermore, the mechanical and thermal stability of polymeric membranes could also be enhanced. The resulting composite membranes were used for dye removal,[48] BSA separation,[67,68] selective separation of monovalent and divalent ions,[69,70] and pervaporation dehydration[48] of different solvent–water mixtures.

6.2.2 Polymer Molecular Modified GO Membrane

Unlike methods of incorporating GO nanosheets into a polymer matrix for improving the membrane performance, GO membrane is also a competitive candidate when directly used as a separating layer in water purification after modification by some polymer molecules to overcome the problem of membrane structure instability. To address the issue of stabilizing the membrane structure without sacrificing permeation, 1,3,5-benzenetricarbonyl trichloride[18] and positively charged poly(allylamine hydrochloride)[19] were employed for cross-linking with GO nanosheets, respectively. The cross-linking process not only tightly interlocked stacked GO nanosheets together to overcome their inherent instability in water but also fine-tuned the charges, functionality and inter-spacing. Moreover, a novel dual-modification strategy, including the cross-linking of GO laminate by EDA and amine-enrichment modification by hyperbranched polyethylenimine, has also been proposed to design stable and highly charged GO membranes for the selective removal of heavy metals *via* nanofiltration (Figure 6.7a).[26] Recently, many more successful works have been carried out by interacting with polymers. Branched polyethylene-imine (BPEI) was intercalated into the stacked GO membrane *via* diffusion during filtration by Jung's group (Figure 6.7b).[71] The GO/BPEI membrane achieved extremely high structure stability, even under sonication for 1 h. Furthermore, BPEI treatment did not affect the filtration performance of the GO film, with the evidence of high rejection (>90%) for the dye molecules methylene blue, rose bengal, and brilliant blue and by their permeation rates of *ca.* 124, 34.8, 12.2, and 5.1%, respectively, relative to those of a typical GO membrane. Also, Xu's group incorporated GO nanosheets with cationic tetrakis(1-methyl-pyridinium-4-yl)porphyrin (TMPyP) to improve structure stability and separation performance in water desalination and purification (Figure 6.7c–d).[72]

In addition to the stability enhancement by polymer molecules, the interlayer spacing within GO membrane was also successfully controlled by chemically cross-linking with diamine monomers, which was inspired by the abundant oxygen-containing groups on the basal plane and edges of GO nanosheets.[43,44] Three diamine monomers, ethylenediamine, butylenediamine, and p-phenylenediamine, were chosen to tune the GO interlayer space from 0.91 to 0.97 and 1.01 nm in the dry state and from 0.93 to 0.99 and 1.09 nm in the wet state, which is slightly different to the values of pristine GO membrane, 0.85 to 1.31 nm (Figure 6.8a).[43] Reduced changes in *d*-spacing from the dry state to the wet state were due to the formed C–N

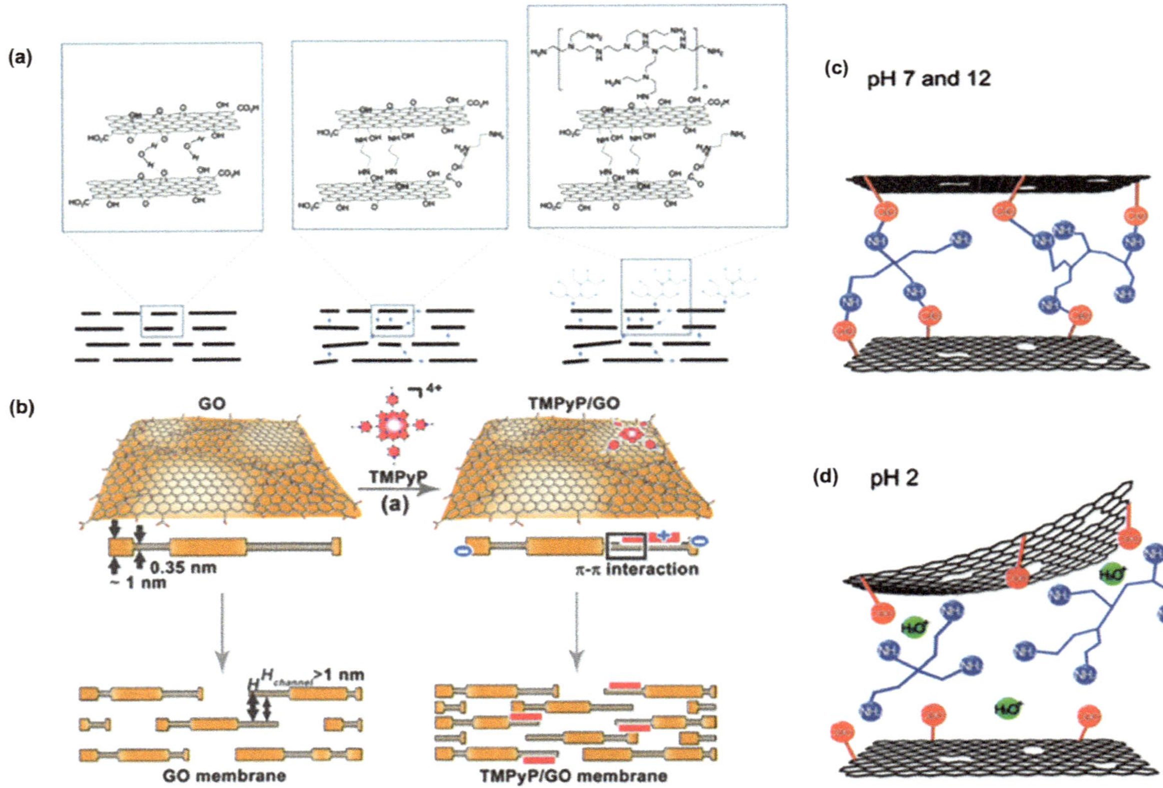

Figure 6.7 (a) Schematic diagrams of the proposed structures: (left) pristine GO, (middle) GO&EDA, and (right) GO&EDA_HPEI 60K frameworks. Reprinted with permission from Y. Zhang, S. Zhang and T. S. Chung, *Environ. Sci. Technol.*, 2015, **49**, 10235, Copyright 2015 American Chemical Society.[26] (b) Assembly of GO with TMPyP; Fabrication of GOLMs Intercalated (left) without or (right) with TMPyP *via* a Vacuum Filtration Process. Reprinted with permission from Y. T. Nam, J. Choi, K. M. Kang, D. W. Kim and H.-T. Jung, *ACS Appl. Mater. Inter.*, 2016, **8**, 27376, Copyright 2016 American Chemical Society.[71] Mechanism of stabilization of the GO/BPEI membrane at different pH values. (c) pH = 7 and 12, and (d) pH = 2. Reprinted with permission from X. L. Xu, F. W. Lin, Y. Du, X. Zhang, J. Wu and Z. K. Xu, *ACS Appl. Mater. Inter.*, 2016, **8**, 12588, Copyright 2016, American Chemical Society.[72]

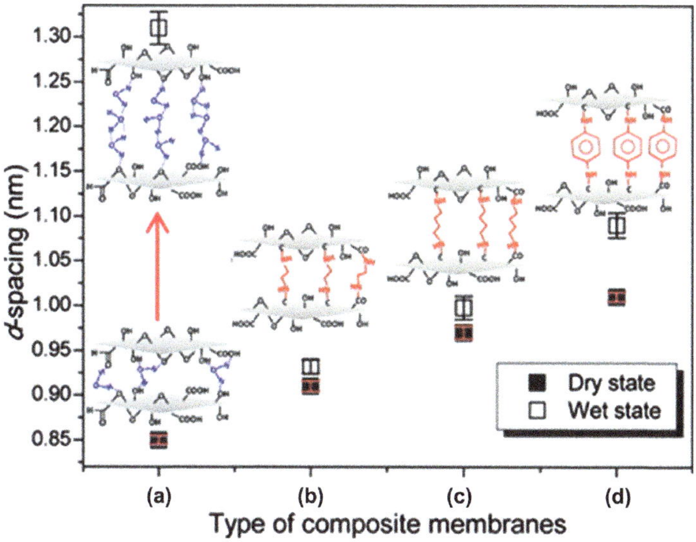

Figure 6.8 Changes in *d*-spacing for composite GO and GOF membranes in dry and wet states. (a) GO/CA; (b) GO–EDA/CA; (c) GO–BDA/CA; (d) GO–PPD/CA. Reprinted with permission from W. S. Hung, C. H. Tsou, M. De Guzman, Q. F. An, Y. L. Liu, Y. M. Zhang, C. C. Hu, K. R. Lee and J. Y. Lai, *Chem. Mater.*, 2014, **26**, 2983, Copyright 2014 American Chemical Society.[43]

covalent bonds, which significantly suppressed the stretching of interlayer spacing. Besides, the cross-liking processes also substantially altered GO hydrophilicity. All of those factors enhanced the pervaporation performance for dehydrating a mixture of alcohol and water. The GO/ethylenediamine membrane presented a short interlayer *d*-spacing and an excellent pervaporation performance at 80 °C (permeation flux: 2297 $g\,m^{-2}\,h^{-1}$, water concentration in permeate: 9.8 *wt* %). Xia *et al.* employed the same strategy with ethylenediamine, 1,3-propanediamine and m-phenylenediamine to control the *d*-spacing for removing natural organic matter using pressure-driven filtration (Figure 6.8b).[44] And, the various ultrathin GO composite membranes exhibited a *d*-spacing induced DOC and UV_{254} removal performance.

In those cases, polymer molecules incorporated into GO membrane not only enhance the structure stability but also fine-tune the charges, functionality and inter-spacing. The obtained excellent selectivity and ultrafast mass transport are derived from the combined effects of nanochannel size, membrane surface conditions (negatively or positively charged, hydrophilicity or hydrophobicity), and functional groups.

6.3 Mass Transport Through Graphene Oxide/ Nanoparticle Composite Membranes

Many successful attempts have been made to add nanoparticles to polymeric or ceramic membranes during membrane synthesis.[73] Now, the same

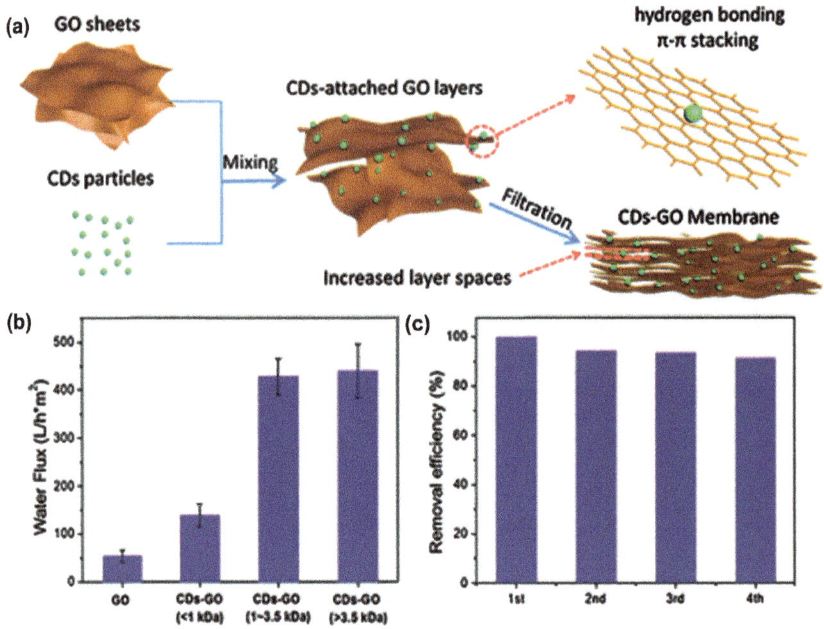

Figure 6.9 Schematic view of possible percolation through the laminate (a) CdS-intercalated GO membranes. (b) The permeability of GO and CD–GO membranes embedded with CDs of different size (note: CD amount is about 2.1%). (c) The cycle tests of CD–GO membranes in MO removal. Reproduced from ref. 75 with permission from the Royal Society of Chemistry.

method has also been extended to GO membranes. Xu *et al.* anchored nano-sized TiO_2 particles onto the surface of GO nanosheets to prop up GO sheets and broaden the interlayer spacing.[74] In comparison to the intrinsic adsorption capacity, the as-prepared GO/TiO_2 composite membranes presented an enhanced organic dye molecular separation performance. A similar strategy was also employed by other groups (Figure 6.9a–c).[75] Wang *et al.* embedded carbon dots with various sizes to create more spaces and channels within GO membranes, causing a sharp enhancement in permeability and without any compromise in the removal of small dye molecules and the rejection of nanoparticles.

6.4 Mass Transport Through Graphene Oxide/Nanowire or Nanotube Composite Membranes

6.4.1 Graphene Oxide/Nanowire Composite Membranes

Recently, Huang *et al.* employed copper hydroxide nanostrands as sacrificial templates to construct a network of nanostrand-channels (NSCs) with a narrow size distribution within GO membranes (Figure 6.10a–c).[40] Highly

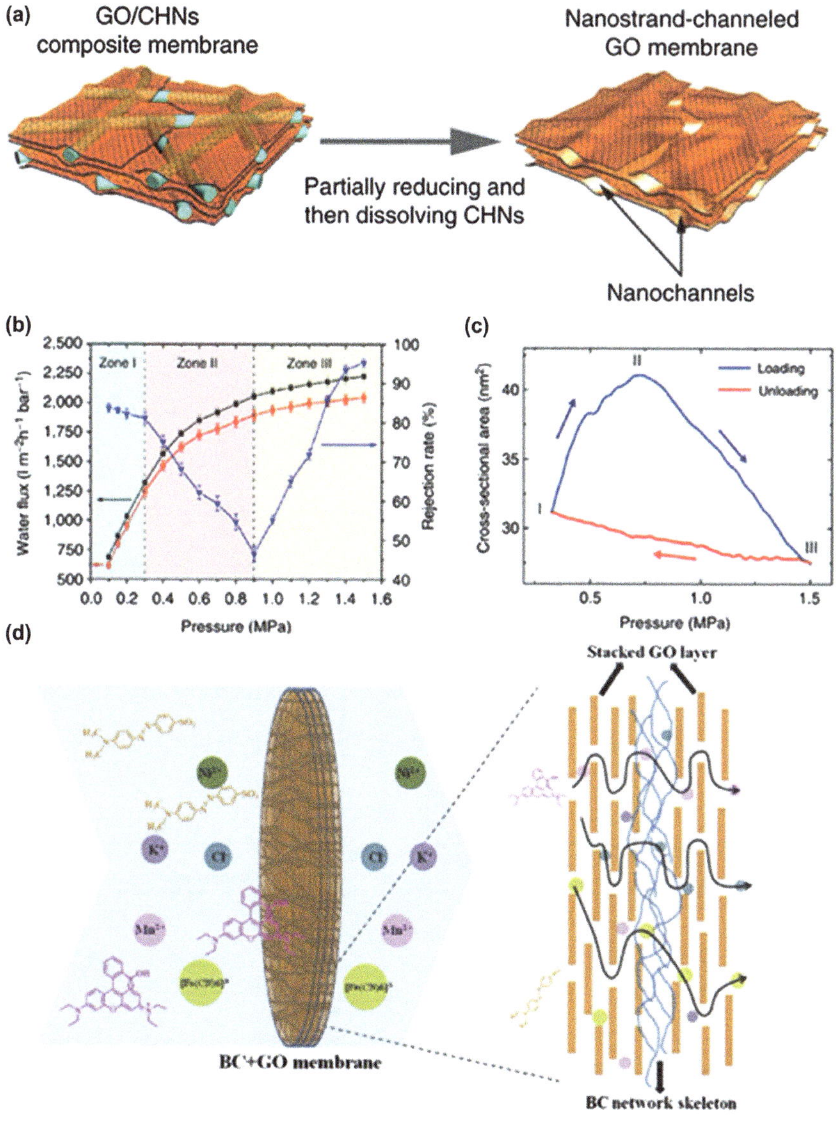

Figure 6.10 (a) Schematic illustration of the nanostrand-channelled GO membrane fabrication process. (b) Water flux and rejection of EB molecules through nanostrand-channelled GO membrane as a function of pressure. The rejection rate refers to the first pressure-loading process. (c) The change in the nanochannel cross-sectional area with respect to the applied pressure obtained by MD simulations. Reprinted by permission from Macmillan Publishers Ltd: *Nature Communications* (ref. 40) Copyright 2013. (d) Schematic of the permeation process of different organic or inorganic ions through BC + GO membrane. Reproduced from ref. 76 under the terms of the CC BY 4.0 licence, https://creativecommons.org/licenses/by/4.0/.

positively charged $Cu(OH)_2$ nanostrands (diameter: 2.5 nm, length: several micrometers, positively charged) can readily form a stable aqueous solution and then tightly incorporate with negatively charged GO nanosheets *via*

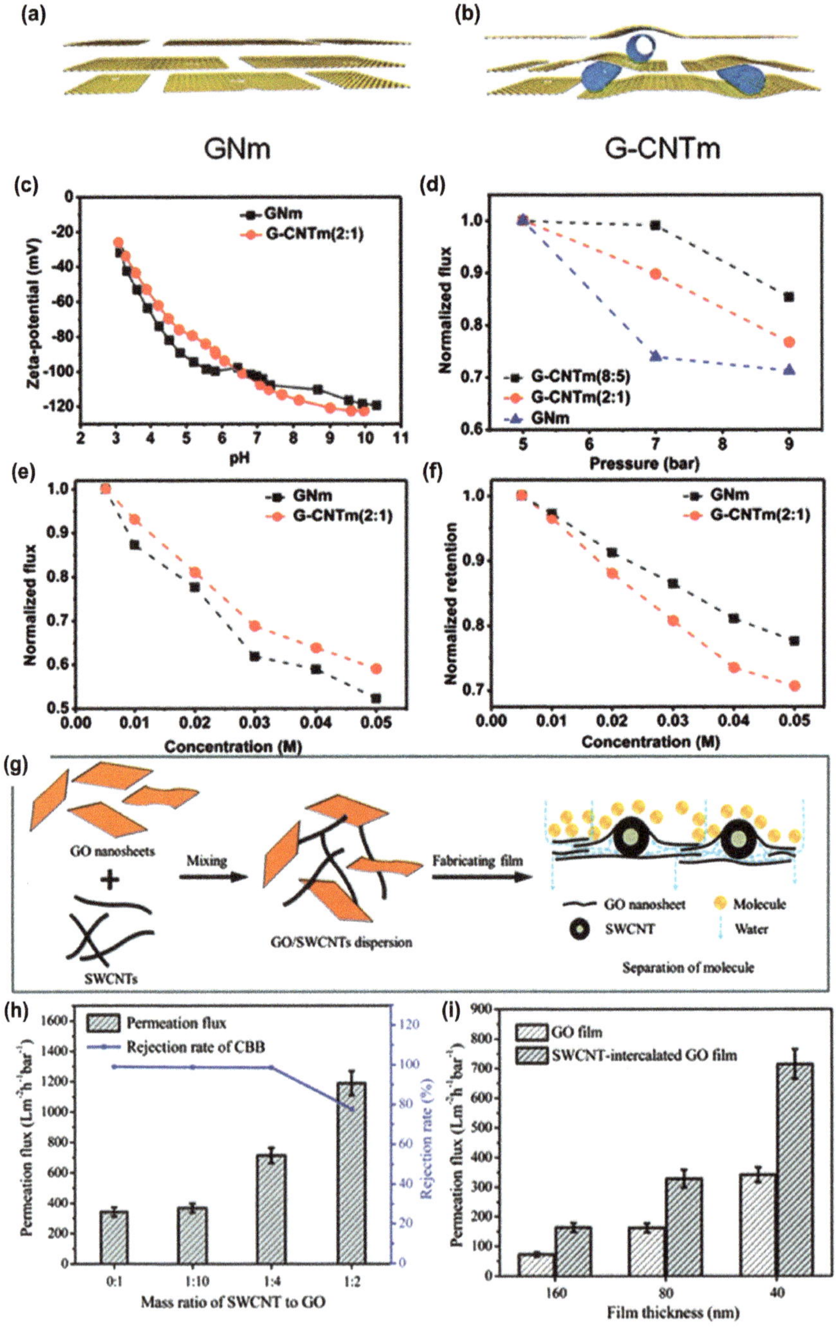

electrostatic interaction. After assembly of NSC–GO membranes on a porous support by filtration and subsequent stabilization of their nanochannel structure by hydrazine treatment, the nanostrands were completely removed by reacting with ethylenediamine tetraacetic acid disodium salt. The constructed nanochannel network dramatically enhanced water permeation, which offered a ten-fold enhancement without sacrificing rejection compared with that of pristine GO membranes, and it was more than 1–2 orders of magnitude higher than that of commercial ultrafiltration membranes with the same rejection. The final separation performance of the as-prepared NSC–GO membranes can be easily controlled by tuning the amount of nanostrands. Remarkably, the elastic deformation of the nanochannels under pressure leads to an abnormal pressure-dependent separation behaviour, demonstrating excellent mechanical properties and offering enormous possibilities to control flux and rejection according to various situations. The results also demonstrated the significance of nanochannels within a membrane.

Recently, a freestanding composite membrane based on bacterial cellulose (BC) and GO was designed and prepared (Figure 6.10d).[76] The BC network provided a porous skeleton to spread the GO nanosheets and uniformly incorporate the GO layers, which endowed the BC/GO composite membrane with good water-stability, excellent tensile strength, and improved toughness, guaranteeing its separation applicability in a water environment. The as-prepared composite membranes presented obviously distinct permeation properties for different inorganic/organic ions of different size, from nanoscale to angstrom-scale. Besides, the possible adsorption of organic dyes (with a benzene ring) onto the GO nanosheets *via* π–π interaction may block the entrance formed by adjacent GO sheets and prevent the access of ions. In this case, the author believes the rejection of organic ions with large size is mainly ascribed to physical sieving by nanochannels rather than π–π adsorption or electrostatic repulsion owing to the small amount of GO covered on the membrane surface and the weak π contribution from GO. Further, the

Figure 6.11 Schematic representation of the structure and water transport path for (a) GNm and (b) G-CNTm. (c) ζ-Potential *versus* pH curves of GNm and G-CNTm(2 : 1). (d) Normalized water flux of GNm and G-CNTms under different driven pressures when filtrating DY (0.02 mM). (e) Normalized water flux and (f) normalized rejection ratio of GNm and G-CNTm(2 : 1) at different salt concentrations when filtrating Na_2SO_4. Reproduced with permission from Y. Han, Y. Q. Jiang and C. Gao, *ACS Appl. Mater. Inter.*, 2015, 7, 8147, Copyright 2015 American Chemical Society.[77] (g) Schematic showing the fabrication process of the SWCNT-intercalated GO ultrathin film and the process of the separation of molecules. (h) Permeation fluxes and rejection rates when separating a CBB solution by the 40 nm-thick SWCNT-intercalated GO films fabricated with different SWCNT/GO mass ratios. (i) Variation in permeation flux as a function of film thickness when separating a CBB solution by using the film with a SWCNT/GO mass ratio of 1 : 4. Reproduced from ref. 42 with permission from Royal Society of Chemistry.

BC network, acting as a supporting skeleton, mainly played the role of maintaining membrane structure stability.

6.4.2 Graphene Oxide/Carbon Nanotube Composite Membranes

Multi-walled CNTs (MWNTs, 50 nm in diameter)[77] and single-walled CNTs (SWCNTs, 5–20 nm in diameter)[42] have also been intercalated into rGO/GO layers to greatly improve water permeation without any sacrifice in rejection.[77] It should be noted that the enhancement in water flux gave solid proof for water transport between graphene layers rather than directly through the pores or pin holes on nanosheets. If not, it would be useless to add MWCNTs into graphene layers (Figure 6.11a–f).[77] But, it seems that the

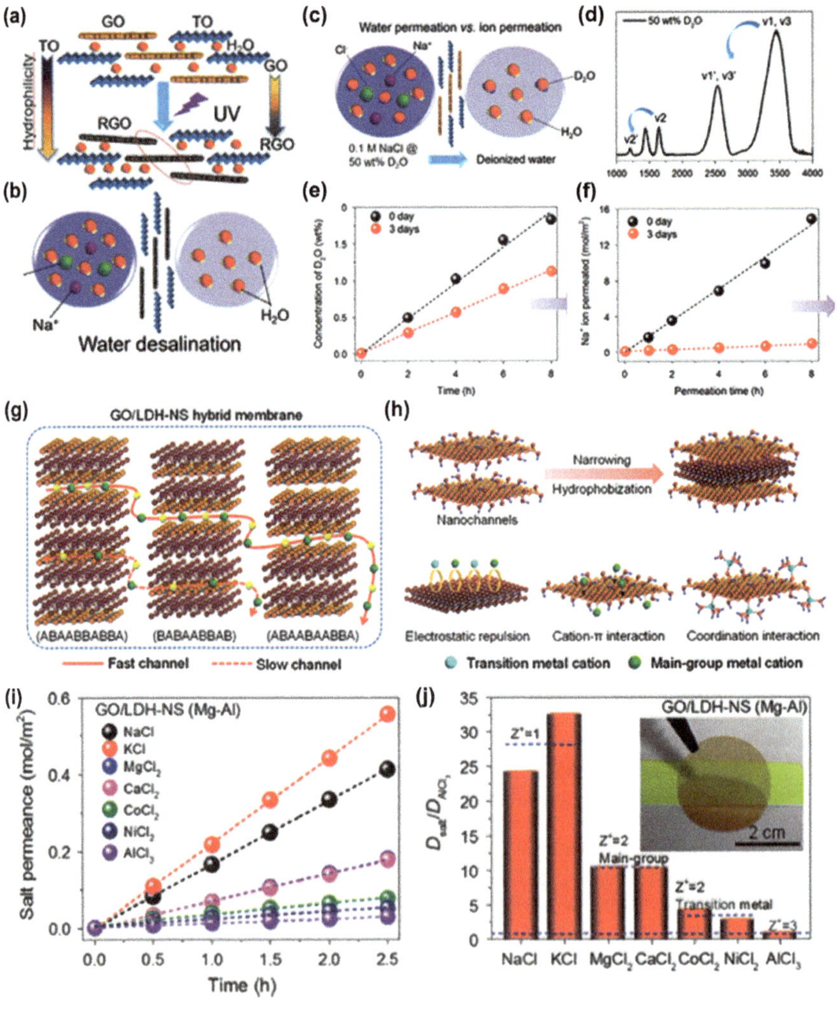

authors ignored the function of intercalated MWCNTs in enlarging the *d*-spacing and increasing the amount of nanochannels (the diameter of the MWCNTs is much larger compared with the size of the original nanochannel and pores or pin holes). Of course, we also believe that water mainly transports between graphene layers. GO/SWCNT composite membranes exhibited excellent performance for the retention of organic dye (>99% for Direct Yellow and >96% for Methyl Orange), especially good rejection of salt ions (*i.e.*, 83.5% for Na_2SO_4, 51.4% for NaCl), and excellent antifouling performance (sodium alginate and humic acid) due to their strong electrostatic interaction and physical sieving properties.[77] Gao *et al.* reported a completely distinct separation performance when intercalating SWCNTs into GO membrane (Figure 6.11g–k). The as-prepared GO/SWCNT composite membranes are superior in the separation of molecules with sizes greater than 1.8 nm, with fluxes of 660–720 $L\,m^{-2}\,h^{-1}\,bar^{-1}$ and separation efficiencies of 97.4% to 98.7% for Bovine Serum Albumin, cytochrome c, Coomassie Brilliant Blue and Rhodamine B.[42]

6.5 Mass Transport Through Graphene Oxide/Other Inorganic Nanosheet Composite Membranes

In additional to zero- and one-dimensional additives within GO membranes, two-dimensional monolayer titania (TO) nanosheets have also been intercalated into GO laminates (Figure 6.12a–f).[78] Under mild UV irradiation, GO nanosheets transformed into rGO due to the photoreduction by TO, which

Figure 6.12 Schematic diagram for the reduction of GO by TO within the hybrid membranes under ultraviolet (UV) irradiation. (a) Photographs of the GO/TO membranes after UV irradiation. (b) Schematic diagram for the water desalination process using RGO/TO hybrid membranes. (c) Schematic diagram for the investigation of ion and water permeations through graphene oxide (GO)(RGO)/titania (TO) membranes based on isotope labelling. (d) Fourier transform infrared spectrum of 50 *wt*% deuterium oxide (D_2O) solution, which indicates the peaks corresponding to H_2O and D_2O. (e) D_2O permeations and the corresponding water diffusion coefficients through GO/TO membranes after ultraviolet (UV) irradiation for 0 and 3 days, respectively. (f) Ion permeations and corresponding diffusion coefficients through GO/TO membranes after UV irradiation for 0 and 3 days, respectively. Reprinted by permission from Macmillan Publishers Ltd: *NPG Asia Materials* (ref. 78) Copyright 2015. (g) Schematic diagram illustrating the possible nanochannels within GO/LDH-NS hybrid membranes. (h) The role of LDH-NS in the modification of GO nanochannels (top panel) and the diverse interactions between various cations and NS (bottom panel). (i) Salt permeations through GO/LDH-NS hybrid membranes (Co-Al) and the corresponding diffusion coefficients (normalized by $AlCl_3$). The insets in (j) are the photographs of GO/LDH-NS hybrid membranes containing Co-Al LDH-NS. Reproduced from ref. 79 under the terms of the CC BY 4.0 licence, https://creativecommons.org/licenses/by/4.0/.

led to a decrease in the interlayer spacing that then resulted in the effective rejection of ions. However, the water permeability was still well-retained, which was attributed to the photoinduced hydrophilic conversion of TO. The special membrane component design and operation conditions endowed the as-prepared rGO/TO hybrid membranes with excellent water desalination performances. Without external hydrostatic pressures, the ion permeations through the rGO/TO hybrid membranes can be reduced to 5% compared with the GO/TO cases, while the water transmembrane permeations can be retained at up to ~60%.

Further, Co–Al (or Mg–Al) layered double hydroxide (LDH) nanosheets (NSs), which are positively charged, were also incorporated with negatively charged GO nanosheets to obtain a hybrid membrane, GO/LDH-NS composite membrane, for high selectivity charge-guided ion transport (Figure 6.12g–j).[79] The interlayer spacing of the hybrid membrane is insensitive to humidity variations, ensuring the structural stability for solution-based mass transport applications. Covering the shortcomings of pure GO membranes, the GO/LDH-NS membranes were highly suitable for engineering-relevant filtration and separation applications, considering the high relative selectivity among cations strictly according to their charges, independent of the specific cation species, and even the slower transport rates of transition metal cations. (The relative selectivity between monovalent and trivalent cations is as high as 30.) Strong narrowing and hydrophobization of the pristine GO nanochannels by the intercalation of LDH-NS, and the synergistic effect of electrostatic interactions from LDH-NS and chemical interactions (coordination, cation–π interaction) from GO-NS were responsible for the effective separation of cations with various valences.

6.6 Conclusion

This chapter shortly addressed recent works on the fabrication of diverse graphene oxide-based composite membranes and the improvement of the mass transport behaviour in separation. The nanochannels within membranes, the surface condition and the functional groups are the three issues of crucial importance for high selectivity and ultrafast flux. On one hand, embedding GO laminates into a polymeric matrix can sharply improve the separation performance. On the other hand, by incorporating GO nanosheets with other materials, such as polymer molecular and nanomaterials from zero- to two dimensional, the as-prepared composite membranes can achieve the desired separation performance though the changes of the three crucial parameters mentioned above. These excellent properties of GO-based composite membranes make them promising candidates in diverse applications, ranging from wastewater treatment and reuse to chemical refinement and biomimetic selective ion transport.

References

1. L. Huang, M. Zhang, C. Li and G. Q. Shi, *J. Phys. Chem. Lett.*, 2015, **6**, 2806.
2. K. A. Mahmoud, B. Mansoor, A. Mansour and M. Khraisheh, *Desalination*, 2015, **356**, 208.
3. Y. Ying, Y. Yang, W. Ying and X. Peng, *Nanotechnology*, 2016, **27**, 332001.
4. K. S. Novoselov, A. K. Geim, S. V. Morozov, D. Jiang, Y. Zhang, S. V. Dubonos, I. V. Grigorieva and A. A. Firsov, *Science*, 2004, **306**, 666.
5. J. C. Meyer, A. K. Geim, M. I. Katsnelson, K. S. Novoselov, T. J. Booth and S. Roth, *Nature*, 2007, **446**, 60.
6. J. S. Bunch, S. S. Verbridge, J. S. Alden, A. M. van der Zande, J. M. Parpia, H. G. Craighead and P. L. McEuen, *Nano Lett.*, 2008, **8**, 2458.
7. S. P. Koenig, L. D. Wang, J. Pellegrino and J. S. Bunch, *Nat. Nanotechnol.*, 2012, **7**, 728.
8. H. B. Huang, Y. L. Ying and X. S. Peng, *J. Mater. Chem. A*, 2014, **2**, 13772.
9. G. P. Liu, W. Q. Jin and N. P. Xu, *Chem. Soc. Rev.*, 2015, **44**, 5016.
10. H. M. Hegab and L. D. Zou, *J. Membr. Sci.*, 2015, **484**, 95.
11. H. Li, Z. N. Song, X. J. Zhang, Y. Huang, S. G. Li, Y. T. Mao, H. J. Ploehn, Y. Bao and M. Yu, *Science*, 2013, **342**, 95.
12. N. F. D. Aba, J. Y. Chong, B. Wang, C. Mattevi and K. Li, *J. Membr. Sci.*, 2015, **484**, 87.
13. X. F. Chen, G. P. Liu, H. Y. Zhang and Y. Q. Fan, *Chin. J. Chem. Eng.*, 2015, **23**, 1102.
14. W. Choi, J. Choi, J. Bang and J. H. Lee, *ACS Appl. Mater. Interfaces*, 2013, **5**, 12510.
15. W. Gao, M. Majumder, L. B. Alemany, T. N. Narayanan, M. A. Ibarra, B. K. Pradhan and P. M. Ajayan, *ACS Appl. Mater. Interfaces.*, 2011, **3**, 1821.
16. K. L. Goh, L. Setiawan, L. Wei, R. M. Si, A. G. Fane, R. Wang and Y. Chen, *J. Membr. Sci.*, 2015, **474**, 244.
17. Y. Han, Z. Xu and C. Gao, *Adv. Funct. Mater.*, 2013, **23**, 3693.
18. M. Hu and B. X. Mi, *Environ. Sci. Technol.*, 2013, **47**, 3715.
19. M. Hu and B. X. Mi, *J. Membr. Sci.*, 2014, **469**, 80.
20. H. B. Huang, Y. Y. Mao, Y. L. Ying, Y. Liu, L. W. Sun and X. S. Peng, *Chem. Commun.*, 2013, **49**, 5963.
21. K. Huang, G. P. Liu, Y. Y. Lou, Z. Y. Dong, J. Shen and W. Q. Jin, *Angew. Chem., Int. Ed.*, 2014, **53**, 6929.
22. R. Joshi, S. Alwarappan, M. Yoshimura, V. Sahajwalla and Y. Nishina, *Appl. Mater. Today*, 2015, **1**, 1.
23. Y. P. Tang, D. R. Paul and T. S. Chung, *J. Membr. Sci.*, 2014, **458**, 199.
24. P. Sun, K. Wang and H. Zhu, *Adv. Mater.*, 2016, **28**, 2287.
25. D. A. Dikin, S. Stankovich, E. J. Zimney, R. D. Piner, G. H. B. Dommett, G. Evmenenko, S. T. Nguyen and R. S. Ruoff, *Nature*, 2007, **448**, 457.
26. Y. Zhang, S. Zhang and T. S. Chung, *Environ. Sci. Technol.*, 2015, **49**, 10235.

27. C. N. Yeh, K. Raidongia, J. J. Shao, Q. H. Yang and J. X. Huang, *Nat. Chem.*, 2015, 7, 166.
28. A. Lerf, A. Buchsteiner, J. Pieper, S. Schottl, I. Dekany, T. Szabo and H. P. Boehm, *J. Phys. Chem. Solids*, 2006, **67**, 1106.
29. K. Raidongia and J. X. Huang, *J. Am. Chem. Soc.*, 2012, **134**, 16528.
30. R. K. Joshi, P. Carbone, F. C. Wang, V. G. Kravets, Y. Su, I. V. Grigorieva, H. A. Wu, A. K. Geim and R. R. Nair, *Science*, 2014, **343**, 752.
31. P. Z. Sun, M. Zhu, K. L. Wang, M. L. Zhong, J. Q. Wei, D. H. Wu, Z. P. Xu and H. W. Zhu, *ACS Nano*, 2013, 7, 428.
32. P. Z. Sun, F. Zheng, M. Zhu, Z. G. Song, K. L. Wang, M. L. Zhong, D. H. Wu, R. B. Little, Z. P. Xu and H. W. Zhu, *ACS Nano*, 2014, **8**, 850.
33. A. Paneri and S. Moghaddam, *Carbon*, 2015, **86**, 245.
34. N. Wei, X. S. Peng and Z. P. Xu, *ACS Appl. Mater. Inter.*, 2014, **6**, 5877.
35. B. X. Mi, *Science*, 2014, **343**, 740.
36. R. R. Nair, H. A. Wu, P. N. Jayaram, I. V. Grigorieva and A. K. Geim, *Science*, 2012, **335**, 442.
37. P. Z. Sun, H. Liu, K. L. Wang, M. L. Zhong, D. H. Wu and H. W. Zhu, *Chem. Commun.*, 2015, **51**, 3251.
38. G. Algara-Siller, O. Lehtinen, F. C. Wang, R. R. Nair, U. Kaiser, H. A. Wu, A. K. Geim and I. V. Grigorieva, *Nature*, 2015, **519**, 443.
39. W. Zhou, K. B. Yin, C. H. Wang, Y. Y. Zhang, T. Xu, A. Borisevich, L. T. Sun, J. C. Idrobo, M. F. Chisholm, S. T. Pantelides, R. F. Klie and A. R. Lupini, *Nature*, 2015, **528**, E1.
40. H. B. Huang, Z. G. Song, N. Wei, L. Shi, Y. Y. Mao, Y. L. Ying, L. W. Sun, Z. P. Xu and X. S. Peng, *Nat. Commun.*, 2013, **4**, 2979.
41. Y. L. Ying, L. W. Sun, Q. Wang, Z. J. Fan and X. S. Peng, *RSC Adv.*, 2014, **4**, 21425.
42. S. J. Gao, H. L. Qin, P. P. Liu and J. Jin, *J. Mater. Chem. A*, 2015, **3**, 6649.
43. W. S. Hung, C. H. Tsou, M. De Guzman, Q. F. An, Y. L. Liu, Y. M. Zhang, C. C. Hu, K. R. Lee and J. Y. Lai, *Chem. Mater.*, 2014, **26**, 2983.
44. S. J. Xia, M. Ni, T. R. Zhu, Y. Zhao and N. N. Li, *Desalination*, 2015, **371**, 78.
45. S. Zinadini, A. A. Zinatizadeh, M. Rahimi, V. Vatanpour and H. Zangeneh, *J. Membr. Sci.*, 2014, **453**, 292.
46. G. Dong, J. Hou, J. Wang, Y. Zhang, V. Chen and J. Liu, *J. Membr. Sci.*, 2016, **520**, 860.
47. J. J. Song, Y. Huang, S. W. Nam, M. Yu, J. Heo, N. Her, J. R. V. Flora and Y. Yoon, *Sep. Purif. Technol.*, 2015, **144**, 162.
48. N. X. Wang, S. L. Ji, G. J. Zhang, J. Li and L. Wang, *Chem. Eng. J.*, 2012, **213**, 318.
49. T. M. Yeh, Z. Wang, D. Mahajan, B. S. Hsiao and B. Chu, *J. Mater. Chem. A*, 2013, **1**, 12998.
50. H. W. Kim, H. W. Yoon, S. M. Yoon, B. M. Yoo, B. K. Ahn, Y. H. Cho, H. J. Shin, H. Yang, U. Paik, S. Kwon, J. Y. Choi and H. B. Park, *Science*, 2013, **342**, 91.
51. W. J. Lau and A. F. Ismail, *Desalination*, 2009, **245**, 321.

52. G. M. Geise, H. S. Lee, D. J. Miller, B. D. Freeman, J. E. Mcgrath and D. R. Paul, *J. Polym. Sci., Polym. Phys.*, 2010, **48**, 1685.
53. P. Marchetti, M. F. J. Solomon, G. Szekely and A. G. Livingston, *Chem. Rev.*, 2014, **114**, 10735.
54. M. M. Pendergast and E. M. V. Hoek, *Energy Environ. Sci.*, 2011, **4**, 1946.
55. J. J. Liang, Y. Huang, L. Zhang, Y. Wang, Y. F. Ma, T. Y. Guo and Y. S. Chen, *Adv. Funct. Mater.*, 2009, **19**, 2297.
56. L. M. Robeson, *J. Membr. Sci.*, 2008, **320**, 390.
57. J. Shen, G. P. Liu, K. Huang, W. Q. Jin, K. R. Lee and N. P. Xu, *Angew. Chem. Int. Ed.*, 2015, **54**, 578.
58. Q. P. Xin, Z. Li, C. D. Li, S. F. Wang, Z. Y. Jiang, H. Wu, Y. Zhang, J. Yang and X. Z. Cao, *J. Mater. Chem. A*, 2015, **3**, 6629.
59. X. Q. Li, Y. D. Cheng, H. Y. Zhang, S. F. Wang, Z. Y. Jiang, R. L. Guo and H. Wu, *ACS Appl. Mater. Inter.*, 2015, **7**, 5528.
60. T. Wang, L. Zhao, J. N. Shen, L. G. Wu and B. Van der Bruggen, *Environ. Sci. Technol.*, 2015, **49**, 8004.
61. Y. Paik, S. S. Kim and O. H. Han, *Angew. Chem. Int. Ed.*, 2008, **47**, 94.
62. H. S. Park, Y. J. Kim, W. H. Hong, Y. S. Choi and H. K. Lee, *Macromolecules*, 2005, **38**, 2289.
63. S. P. Jiang, Z. Liu and Z. Q. Tian, *Adv. Mater.*, 2006, **18**, 1068.
64. B. G. Choi, Y. S. Huh, Y. C. Park, D. H. Jung, W. H. Hong and H. Park, *Carbon*, 2012, **50**, 5395.
65. A. Enotiadis, K. Angjeli, N. Baldino, I. Nicotera and D. Gournis, *Small*, 2012, **8**, 3338.
66. Y. Heo, H. Im and J. Kim, *J. Membr. Sci.*, 2013, **425**, 11.
67. Z. H. Wang, H. R. Yu, J. F. Xia, F. F. Zhang, F. Li, Y. Z. Xia and Y. H. Li, *Desalination*, 2012, **299**, 50.
68. Y. F. Zhao, Z. W. Xu, M. J. Shan, C. Y. Min, B. M. Zhou, Y. L. Li, B. D. Li, L. S. Liu and X. M. Qian, *Sep. Purif. Technol.*, 2013, **103**, 78.
69. B. M. Ganesh, A. M. Isloor and A. F. Ismail, *Desalination*, 2013, **313**, 199.
70. J. Yin, G. Zhu and B. Deng, *Desalination*, 2016, **379**, 93.
71. Y. T. Nam, J. Choi, K. M. Kang, D. W. Kim and H.-T. Jung, *ACS Appl. Mater. Inter.*, 2016, **8**, 27376.
72. X. L. Xu, F. W. Lin, Y. Du, X. Zhang, J. Wu and Z. K. Xu, *ACS Appl. Mater. Inter.*, 2016, **8**, 12588.
73. M. Baghbanzadeh, D. Rana, C. Q. Lan and T. Matsuura, *Sep. Purif. Rev.*, 2016, **45**, 141.
74. C. Xu, A. J. Cui, Y. L. Xu and X. Z. Fu, *Carbon*, 2013, **62**, 465.
75. W. T. Wang, E. Eftekhari, G. S. Zhu, X. W. Zhang, Z. F. Yan and Q. Li, *Chem. Commun.*, 2014, **50**, 13089.
76. Q. Fang, X. Zhou, W. Deng, Z. Zheng and Z. Liu, *Sci. Rep.*, 2016, **6**, 33185.
77. Y. Han, Y. Q. Jiang and C. Gao, *ACS Appl. Mater. Inter.*, 2015, **7**, 8147.
78. P. Z. Sun, Q. Chen, X. D. Li, H. Liu, K. L. Wang, M. L. Zhong, J. Q. Wei, D. H. Wu, R. Z. Ma, T. Sasaki and H. W. Zhu, *NPG Asia Mater.*, 2015, **7**, e162.
79. P. Sun, R. Ma, W. Ma, J. Wu, K. Wang, T. Sasaki and H. Zhu, *NPG Asia Mater.*, 2016, **8**, e259.

CHAPTER 7

Graphene-based Membranes for Barrier Applications

SHANSHAN CHEN,*[a] KAI-GE ZHOU[b] AND HAO YING[c]

[a] Renmin University of China, Department of Physics, Beijing 100872, China; [b] The University of Manchester, National Graphene Institute, Manchester, M13 9PL, UK; [c] Xiamen University, Department of Physics, Xiamen 361005, China
*Email: schen@ruc.edu.cn

7.1 Introduction

Graphene is composed of sp^2 hybridized carbon atoms linked to one another in a 2D honeycomb lattice with high electron-density in its aromatic rings, blocking-off all molecules.[1] Besides its complete impermeability, graphene is also known for its remarkable properties such as high mechanical strength (Young's modulus of 1 TPa),[2] high thermal conductivity,[3] chemical stability,[4] and optical transparency.[5] All of these properties of graphene exhibit its great potential as a barrier membrane. In this chapter, we review the fundamental physical properties of graphene and its derivatives in barrier applications. The latest developments in the properties of barrier membranes by using graphene-based materials, including perfect graphene lattice, chemical vapor depositied (CVD) graphene, graphene oxide (GO) and their composites, are reviewed here in relation to their potential applications. At the end, a summary and outlook is provided on the opportunities and challenges in this rising field.

Nanoscience & Nanotechnology Series No. 47
Graphene-based Membranes for Mass Transport Applications
Edited by Pengzhan Sun and Hongwei Zhu

Published by the Royal Society of Chemistry, www.rsc.org

7.2 Barrier Membranes Based on Perfect Graphene Lattice

The π-orbitals of perfect graphene lattice form a dense and delocalized cloud within its aromatic rings, which does not allow the penetration of even the smallest atoms or molecules.[1,6] As shown in Figure 7.1, the calculated geometric pore gap of the carbon ring is only 0.064 nm,[7] much smaller than the kinetic diameter of the smallest He gas, that is, 0.26 nm. This is further supported by the study using the first-principles density functional theory by Leenaerts *et al.*, which examined the penetration of a helium atom through the centre of a carbon hexagon of a perfect graphene monolayer.[8] It was found that the energy barrier for a helium molecule to approach the centre of a carbon ring was calculated to be 18.8 eV, higher than the kinetic energy of the helium atom (18.6 eV). Defective graphene layers showed an exponentially decreasing penetration energy barrier with the size of the defects, but it was still relatively high enough that it required very large defects to permeate small atoms and molecules through graphene sheets.

In 2008, Bunch *et al.* performed the first experimental study to demonstrate the impermeability of the non-defective, mechanically exfoliated graphene.[6] As shown in Figure 7.2a–b, the permeance of several gases through the micro chamber capped with a graphene sheet was tested by measuring the gas leak rates. It was found that the gas leak rate was independent of the thickness of the graphene membrane, indicating that the leak was not through the graphene sheets. These results demonstrate that perfect graphene is the thinnest barrier for all the molecules in the world. Besides

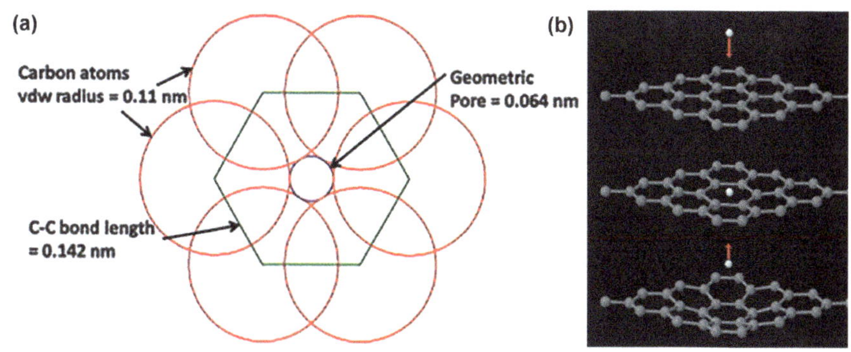

Figure 7.1 (a) Molecular structure of graphene with a very small geometric pore showing its impermeability to all molecules at room temperature. Reprinted from *Carbon*, **62**, V. Berry, Impermeability of graphene and its applications, 1–10, Copyright 2013, with permission from Elsevier.[7] (b) Reflection of a He atom with a kinetic energy of 18.6 eV from a graphene surface: The He atom approaches the perfect graphene layer, then comes to rest before penetrating the graphene layer, and finally it is reflected back. Reprinted from O. Leenaerts, B. Partoens and F. Peeters, *ApPhL*, 2008, **93**(19), 193107, with the permission of AIP Publishing.[8]

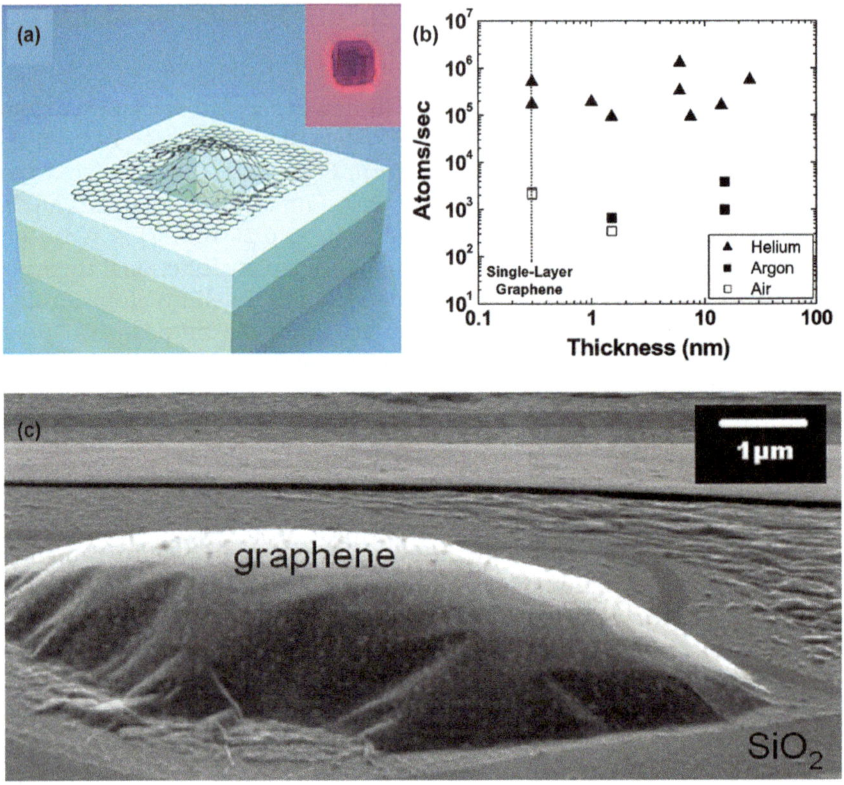

Figure 7.2 (a) Schematic of a mechanically exfoliated graphene sealed microchamber. The inset shows an optical image of a monolayer graphene drumhead on SiO_2. (b) Gas leakage *vs.* membrane thickness for all the devices measured. Reprinted with permission from J. S. Bunch, S. S. Verbridge, J. S. Alden, A. M. Van Der Zande, J. M. Parpia, H. G. Craighead and P. L. McEuen. *Nano Lett.*, 2008, **8**(8), 2458, Copyright 2008 American Chemical Society.[6] (c) SEM image of a single layer graphene flake after exposure to HF vapour, taken at a 75° angle with respect to the sample plane. Reprinted with permission from E. Stolyarova, D. Stolyarov, K. Bolotin, S. Ryu, L. Liu, K. Rim, M. Klima, M. Hybertsen, I. Pogorelsky and I. Pavlishin. *Nano Lett.*, 2008, **9**(1), 332, Copyright 2009 American Chemical Society.[4]

impermeability, graphene barrier has shown remarkable chemical inertness to most of the gases and fluids (even to the most aggressive chemicals such as HF). Mechanically exfoliated graphene was mounted on a SiO_2/Si substrate that was subjected to HF/H_2O etching.[4] The released gases from SiO_2 can be captured under one atom thick graphene sheets for months without showing signs of degradation, demonstrating the robust nature of the graphene sheets (Figure 7.2c).

Recently, Hu *et al.* reported that monolayers of graphene are highly permeable to protons, even at ambient conditions, whereas no proton transport was detected for thicker crystals such as the bilayer graphene.[9] This can be attributed to the existence of "pores" in the monolayer graphene's electron clouds, which become much denser for bilayer graphene. This tuneable proton conductivity combined with chemical and thermal stability, and impermeability makes the graphene membrane an attractive candidate to be used in various hydrogen technologies.

Single-crystal graphene with a perfect lattice has shown great potential as an excellent barrier membrane. However, the mechanical exfoliation technique is not suitable for practical industrial applications because of the low yield and small size of the graphene sheets produced. Therefore, it would be more practical to utilize CVD graphene, GO, rGO and other graphene derivatives for designing graphene-based membranes for various barrier applications.

7.3 Barrier Applications Based on CVD Graphene Membranes

CVD and epitaxial growth are suitable techniques for producing high-quality, large area graphene. In particular, CVD techniques have been successfully applied to grow high-quality single-layer and multilayer graphene on various metals[10–13] or even insulating substrates.[14–16] However, CVD graphene is typically polycrystalline with many intrinsic defects (~1–15 nm in size) or grain boundaries that may allow unfavourable molecular transport.[17–19] Despite the imperfect graphene lattice, a number of studies have used CVD graphene as anticorrosion layers, metal diffusion barriers in microelectronics, wear and friction resistant layers[20,21] and as a protective film in biological environments.[22]

7.3.1 Antioxidation and Anticorrosion Barriers

Since modern civilization started using artificial materials such as steel and plastic, their corruption caused the annual spending of trillions of dollars deterring faster incorporation of modern technology. In the battle of Tsushima strait, the speed of the Russian Baltic fleet was about 8 knots slower than the Japanese fleet, due to their heavily fouled hulls. The Russian fleet was completely annihilated in two days, and the consequence of this battle influenced the 20th century of the Pacific Rim. Nowadays, trillions of dollars are wasted every year across the world due to corrosion issues. In a typical process of corrosion, a metal is firstly dissolved into moisture, leaving negative charges in the main body of the metal. Then, the negative charges move to the surface and reduce O_2 in the atmosphere as the surface of the metal is oxidized. High salinity in the environment can speed up this rusting process. Therefore, an ideal anticorrosion barrier should be able to isolate the metal from oxygen, moisture and salt. Traditional approaches to protect the surface of reactive metals or alloys from corrosion include coating with organic layers[23] or polymers,[24] formation of oxide layers,[25] anodization,[26] chemical modification,[27] and coating with other metals or alloys.[28] The addition of a protective coating changes the dimensions of the metal due to the finite thickness of the coating, changes the appearance and the optical properties of the metal surface, and often decreases the electrical and thermal conductivity. Therefore, the development of new materials for anticorrosion is in high demand.

The full potential of graphene as a novel protective anticorrosion barrier can be understood based on its impermeability, chemical stability, and other physical properties (optical transparency, thermal stability *etc.*). In particular, monolayer or multilayer graphene could be directly grown on various metallic and insulating substrates or mechanically transferred onto arbitrary surfaces. Pioneering studies have shown that the CVD grown graphene can inhibit the oxidation of the underlying Cu or Cu/Ni alloy metals in air at high temperature or in H_2O_2 solutions (Figure 7.3a).[29] It was also reported that better protection could be afforded by a multilayer graphene, since the defect sites or the graphene grain boundaries of the monolayer graphene surface are more easily attacked (Figure 7.3b). Prasai *et al.* reported that the electrochemical response of graphene-coated copper is substantially decreased with a corrosion rate between 85 and 95%.[30] Moreover, it has been observed that Ni with the transferred multilayer graphene on its surface corrodes 20 times slower than bare nickel (Figure 7.3c–d).

Given the huge cost caused by corrosion, a mitigation of corrosion of metals through graphene coating is immensely attractive. However, some long-term studies found that the corrosion resistance of graphene coating on Cu drastically deteriorated due to the cathodic character of graphene toward Cu in the presence of electrolytes.[31–33] Schriver *et al.* showed that O_2 and H_2O can infiltrate through defects in the graphene and oxidize or corrode the surface of the underlying substrate over long timescales (months or longer).[31] Conducting graphene coatings can further promote corrosion by facilitating electrochemical reactions both across the surface and through the bulk of the substrate (Figure 7.4a–b), whereas typical native oxides passivate the surface and terminate the electron transfer needed for continued corrosion. In light of these phenomena, the origin of the incomplete passivation has to be found and strategies for perfect passivation have to be identified. Hsieh *et al.* used a combined approach of electrochemical analysis and morphological characterization to study the origin of the limited passivation effect of the Cu/graphene system.[32] They found that nanometre-sized structural defects in the graphene are responsible for the limited passivation effect. Through selective passivation of the defects by atomic layer deposition (ALD), an enhanced corrosion protection of more than 99% was achieved, making the Cu/graphene/ALD system one of the best corrosion inhibitors reported (Figure 7.4c–d).

Polycrystalline graphene grown by CVD contained defective grain boundaries and some intrinsic defects that could have allowed the penetration of hazardous or corrosive molecules into the protected substrate

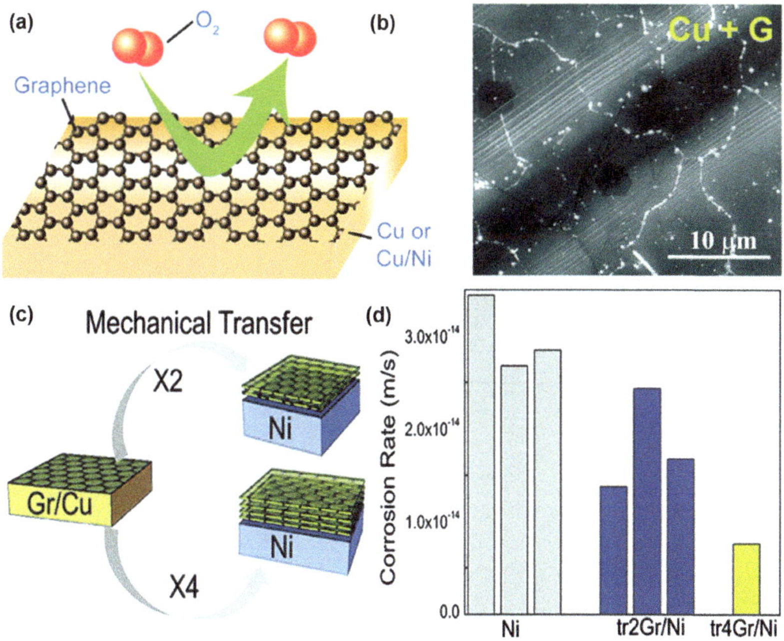

Figure 7.3 (a) Illustration of a graphene sheet as a chemically inert diffusion barrier. (b) SEM image of graphene coated Cu foil taken after annealing in air. Reprinted with permission from S. Chen, L. Brown, M. Levendorf, W. Cai, S. Y. Ju, J. Edgeworth, X. Li, C. W. Magnuson, A. Velamakanni and R. D. Piner. *ACS Nano*, 2010, **5**(2), 1321, Copyright 2011 American Chemical Society.[29] (c) Graphene on Ni sample grown *via* CVD and transferred 2 and 4 layer graphene onto Ni surfaces. (d) Comparison of the corrosion rates of bare Ni samples and the graphene protected Ni samples. Reprinted with permission from D. Prasai, J. C. Tuberquia, R. R. Harl, G. K. Jennings and K. I. Bolotin. *ACS Nano*, 2012, **6**(2), 1102, Copyright 2012 American Chemical Society.[30]

surface. To make CVD graphene films more protective, less defective and large-grain-sized graphene should be synthesized and transfer techniques should be improved. On the other hand, new techniques to passivate the defects in graphene sheets could be another way to enhance the performance of graphene barrier for corrosion inhibition.

7.3.2 Barrier Applications in Microelectronics

Mitigating the interdiffusion of contact metals or the reaction of metals with the underlying semiconductor during thermal processing and device operation is important, especially for ohmic contacts whose resistance can be severely compromised. Protective layers are required to minimize the interfacial diffusion. There are several key properties required for solid-state diffusion barriers in microelectronics, including thermodynamic stability between layers, electrical conductivity, strong adhesion, low contact

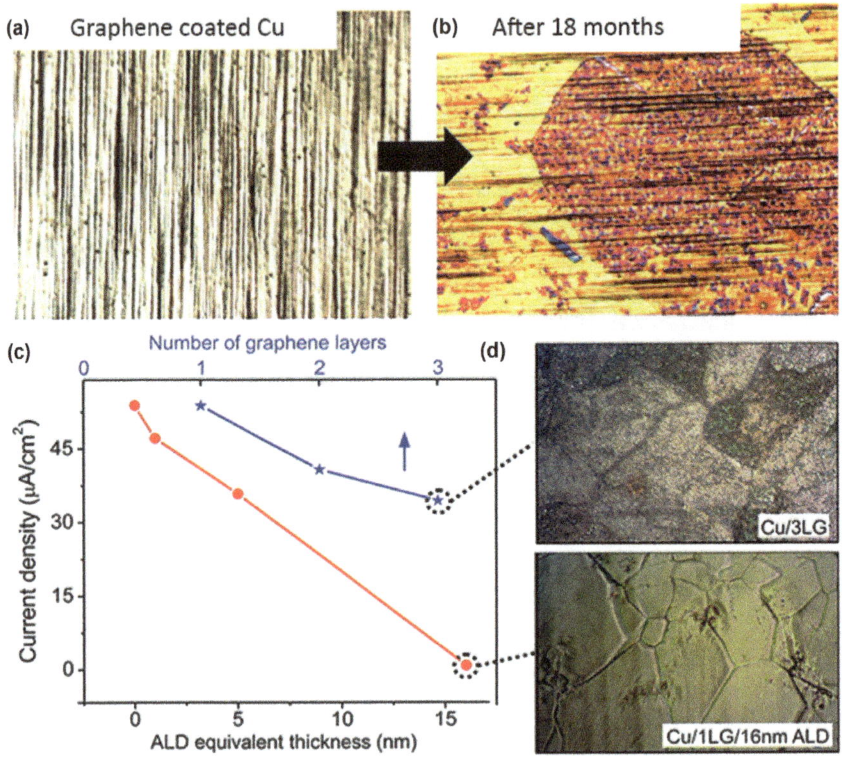

Figure 7.4 (a) Graphene-coated Cu immediately after growth. (b) Graphene-covered Cu 18 months after graphene growth. The entire Cu surface now appears corroded. Reprinted with permission from M. Schriver, W. Regan, W. J. Gannett, A. M. Zaniewski, M. F. Crommie and A. Zettl. *ACS Nano*, 2013, 7(7), 5763, Copyright 2013 American Chemical Society.[31] (c) Peak current of Cu^0 peak for Cu/graphene/ALD samples with different ALD thickness and comparison to peak current for Cu/graphene samples with 1–3 layers of graphene. (d) Comparison of changing Cu morphology by etching using different passivation methods after CV corrosion for Cu/3LG and Cu/1LG/16 nm ALD. Reprinted with permission from Y.-P. Hsieh, M. Hofmann, K.-W. Chang, J. G. Jhu, Y.-Y. Li, K. Y. Chen, C. C. Yang, W.-S. Chang and L.-C. Chen. *ACS Nano*, 2014, **8**(1), 443, Copyright 2014 American Chemical Society.[32]

resistance, and mechanical and thermal resistance. Graphene, with its exceptional physical and chemical properties, has attracted high interest for applications as an effective diffusion barrier for metals on semiconductors.

The insertion of CVD graphene layers between Al metal layer and Si substrates (Figure 7.5b) was reported to provide a significant reduction of spiking and intermixing between the metal contacts and the Si layer.[34] The inserted graphene layer prevented reactions up to 700 and 600 °C at the Al/Si interfaces. A cross-sectional scanning electron microscopy (SEM) view of Al spikes formed in the Al/Si interface annealed at 600 °C is shown in

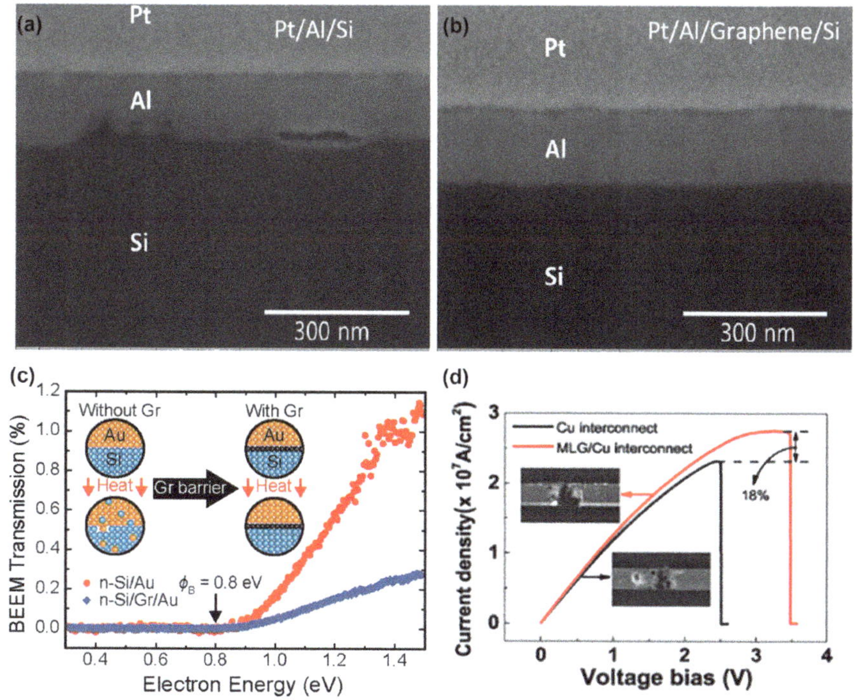

Figure 7.5 (a–b) Cross sectional scanning electron microscopy (SEM) images of (a) Pt/Al/Si and (b) Pt/Al/graphene/Si structures after annealing at 600 °C. Reprinted from H.-Y. Kim, C. Lee, J. Kim, F. Ren and S. Pearton. *J. Vac. Sci. Technol. B*, 2012, **30**(3), 030602, with the permission of AIP Publishing.[34] (c) BEEM spectra obtained for the n-Si(111)/Au (15 nm) and the n-Si(111)/Gr/Au (15 nm) interfaces. Reprinted with permission from C. P. Y. Wong, T. J. H. Koek, Y. Liu, K. P. Loh, K. E. J. Goh, C. Troadec and C. A. Nijhuis. *ACS. Appl. Mat. Inter.*, 2014, **6**(22), 20464, Copyright 2014 American Chemical Society.[36] (d) Characterization of breakdown points of Cu and 5 nm MLG/Cu interconnects. The inset shows the SEM photographs of the interconnect breakdown points. Reproduced from ref. 37 with permission from IOP Piblishing.[37]

Figure 7.5a–b. The thermal stability of graphene is very useful to improve the stability of the metallization at elevated temperatures. The insertion of a graphene layer between a semiconducting indium gallium zinc oxide layer and the Al metal offered good stability during repeated annealing cycles at high temperature while maintaining ohmic contact.[35] Moreover, Wong *et al.*[36] studied the electronic properties of the n-Si(111)/Graphene/M Schottky barriers (M = Au or Cu) by I(v) measurements at the nanoscale by using ballistic electron emission spectroscopy (BEEM).[36] It was shown that the graphene layer does not change the Schottky barrier of Si/Au or Si/Cu junctions, and it effectively prevents diffusion of the Cu or Au into the Si at elevated temperatures (Figure 7.5c).

Copper is widely used as an interconnect in modern ICs because of its low electrical resistivity and good electromigration reliability. However, Cu atoms readily diffuse into Si and SiO_2 even at temperatures as low as 200 °C, forming resistive Cu–Si compounds and generating deep trap levels in the silicon. This results in degradation of the device performance and reliability. The evolution of copper-based interconnects requires the realization of an ultrathin diffusion barrier layer between the Cu interconnect and insulating layers. In recent years, the benefits of graphene as the Cu diffusion barrier have been extensively studied.[37–41] Transferred multilayer graphene (MLG) capping on Cu interconnects has been experimentally demonstrated.[37] The resistance of MLG capped Cu wires improved by 2–7% compared to Cu wires. The breakdown current density increased by 18%, suggesting that the MLG can act as an excellent capping material for Cu interconnects, improving the reliability (Figure 7.5d). However, Bong *et al.* found that even though single-layer CVD graphene exhibited excellent diffusion barrier performance for Cu interconnects, it had poor integration compatibility, displaying weak adhesion and poor nucleation for deposition on top of Cu.[41] The use of an extremely thin reduced graphene oxide (rGO) layer as a Cu barrier material is expected to provide an alternative route for further scaling of copper interconnect technology.

As a solid-state diffusion barrier, CVD graphene has proved to be a promising candidate. The adhesion between graphene and the interface is of vital importance for the stability and performance of devices. Therefore, continued research to improve bonding techniques can further advance its application in microelectronics.

7.4 Barrier Applications of GO, rGO and their Composites

GO, a graphene with rich oxygen content, is another excellent graphene derivative with great promise for mass production.[42,43] GO consists of oxygen-containing functional groups on the basal plane, and it can be well-dispersed in polar solvents such as water. The GO flakes in water can be easily reassembled into a macroscopic lamellar membrane *via* various methods such as drop-casting,[44] vacuum filtration,[45] spin-coating,[46] and so on. The resulting GO laminates possess excellent mechanical strength due to the strong interlayer hydrogen bonds between the adjacent GO layers that hold all the flakes tightly. Consequently, as-prepared GO membranes can be detached from the substrates for freestanding operation. In addition, through chemical treatment, epoxy groups in GO can be removed to produce reduced graphene oxide (rGO) and the decreased conductivity in the nanosheets due to oxidation can be partially restored. On the basis of the above advantages, GO and its reduced form rGO membranes are believed to be the most promising graphene derivative for practical barrier applications.[43]

7.4.1 Gas Barriers

Gas barrier membranes have been applied in many packaging applications, such as in food preservation, pharmaceuticals, and encapsulation of electronic devices. An ideal gas barrier membrane should have ultralow gas permeability and high enough mechanical strength to survive in practical applications. Among different applications, organic light emitting diodes (OLEDs) demand the strictest requirements in the properties of gas barrier membranes. This is due to the susceptibility of the organic molecules in OLEDs to oxygen and moisture, especially when the device is exposed to light with external voltage causing the excitation of the luminescent aromatic molecules. At present, LED displays use glass or metal as an encapsulation layer. However, one of the main advantages of OLEDs is their flexibility, which can be used for flexible applications such as wearable personal devices. Therefore, the most ideal gas barrier membrane should be both flexible and transparent with high impermeability to block the gas diffusion.

The gas permeation rate (Q) through a membrane can be expressed as:

$$Q = DSAt(\Delta p/t_m) \tag{7.1}$$

where D and S are the diffusion and solubility coefficients of the gas through the membrane, respectively; A represents the effective area of the membrane; t is the mobile time of the gas; Δp is the pressure difference across the layer; and t_m is the effective thickness of the membrane when both sides of the membrane are under vacuum. Except for the diffusion (D) and solubility (S) coefficients, the other parameters are dependent on the engineering of the membrane fabrication process. Therefore, the gas permeability of the membrane (P) is defined as:

$$P = DS \tag{7.2}$$

The above equation is often used to describe the gas permeation when the diffusion of gas is irrelevant to the concentration and the solubility coefficient obeys Henry's law. If the membrane is constructed of two components, permeable and impermeable, the solubility coefficient will be modified as:

$$S = S_0(1 - \varphi) \tag{7.3}$$

where φ represents the volume fraction of the impermeable component, and S_0 is the solubility of the permeable component. When the gas diffusion pathway becomes tortuous, the diffusion coefficient will be changed as follows:

$$D = D_0/\tau \tag{7.4}$$

where τ represents the tortuosity of the membrane, and D_0 is the diffusion coefficient of the permeable component. τ can be expressed as:

$$\tau = t_m/t'_m \tag{7.5}$$

Table 7.1 Representative gas barriers based on GO/rGO and their composites.

Membrane	Processing	Gas	Permeation rate	Gas barrier	Ref.
GO-0.5μm	Filtration	H_2O	10^{-5} mm g cm^{-2} s^{-1} bar^{-1}	No	47
GO-5 nm	Spin-coating	O_2	10GPU	No	54
rGO-0.5μm	Filtration/HI Reduction	H_2O	$<10^{-11}$ mm cm^3 s^{-1} bar^{-1} cm^{-2}	Yes	50
$(PEI/GO)_{20}$-98 nm	LbL Assembly	O_2	2.5×10^{-20} cm^3 cm cm^{-2} s^{-1} Pa^{-1}	Yes	64
PLA + 1.37wt%GO	Mixing	O_2	1.1×10^{-14} cm^3 cm cm^{-2} s^{-1} Pa^{-1}	Yes	59
PI + 5wt%GO	Grafting	O_2	0.15×10^{-13} mol s^{-1} m^{-1} atm^{-1}	Yes	63
PS + 0.05vol%rGO	Grafting	O_2	1.84 ± 0.02 Barrer	Yes	60
PP + 1wt%rGO	Mixing	O_2	58.2 mol m^{-2} s^{-1}	Yes	58

where t'_m is the shortest pathway to diffuse through the membrane. Therefore, the two main strategies to improve the performance of the gas barrier membranes are to increase the volume fraction of the impermeable component and the tortuosity of the membrane.

The rise of graphene showed great potential for accelerating the development of gas barrier membranes. GO, rGO and their composites have been widely used in gas barriers (summarized in Table 7.1). Due to the chemical oxidation process, few nanometre pores are observed in the plane of GO or rGO.[44] As a result, the single layered GO or rGO cannot block most gases, *e.g.*, O_2. A laminate film of GO from several micrometres[47] to a few nanometers[48,49] is an alternative candidate for gas barrier membranes. In the laminate of GO, the mass transport takes place along the interlayer spacing, ranging from 0.7 to 1 nm. The vacancy-free GO/rGO plane can be treated as a gas impermeable area. Therefore, the terminal parameter is the tortuosity of the GO membrane. For the laminate structure (Figure 7.6a), the length of diffusion path t'_m can be expressed as:

$$t'_m = t_m + \langle N\rangle L/2\langle N\rangle = t_m/(D'+W) \tag{7.6}$$

where L and W represent the length and thickness of an individual GO sheet, and $\langle N\rangle$ denotes the average number of nanosheets across the membrane. In a GO laminate, W is only a few atoms thick, while L is extended to hundreds of nanometres. Consequently, the tortuosity, τ, of GO laminate is far beyond that of other membrane materials.

In principle, helium can be blocked by the interlayer confinement effect, due to the interlocked hydrogen bonding. However, water molecules can intercalate into the interlocked hydrogen bonding network causing swelling of the interlayer spacing and making the laminate permeable to moisture (Figure 7.6b).[53] Consequently, pristine GO membrane is not suitable as a moisture/O_2 gas barrier.[54] However, the properties of the membrane can be significantly improved to block moisture by chemical reduction in order to remove the interlocked hydrogen bonding network. Su *et al.* compared different chemical reduction methods to improve the gas barrier performance

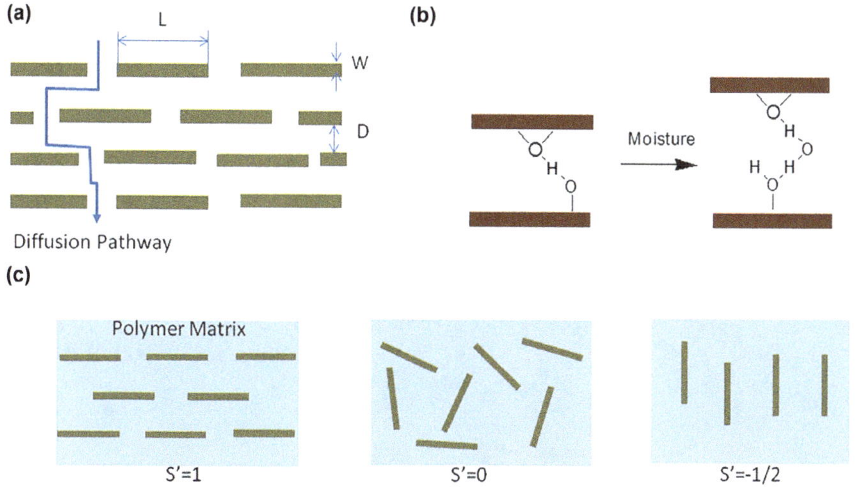

Figure 7.6 (a) Gas diffusion model in the laminate structure; (b) moisture-induced swelling of GO laminate; (c) gas diffusion model in a polymer composition with GO/rGO nanosheets.

of rGO (Figure 7.7a).[50] They found that ascorbic acid or hydroiodic acid treatment can reduce the moisture permeation of rGO (>200 nm) below 10^{-9} mm cm^3 s^{-1} bar^{-1} cm^{-2}, showing better barrier performance than a 30 nm alumina coating.[50] It should be noted that a gas-free reduction technique should be employed for the fabrication process of rGO. This is because the release of gas during the reduction process, such as in the case of thermal reduction, can disturb the laminate structure and therefore the gas barrier properties cannot be improved.[50]

Combining GO/rGO with a polymer matrix is another way to improve the gas barrier properties.[55] The main body of the composite is polymer, and the GO/rGO acts as filler. By loading a few percentage of GO/rGO, the gas barrier properties of polymer composites were dramatically enhanced.[55,56] GO or rGO acts as a "barricade" on the pathway of gas diffusion to increase the tortuosity of the membrane (Figure 7.6c), so the small molecules have to take roundabout routes. Consequently, the length of diffusion in the composite membrane is extended, while the thickness of the barrier membrane remains the same. The permeability of GO/polymer matrix can be modified from eqn (7.2), (7.3) and (7.4) as:

$$P = P_0(1 - \varphi)/\tau \tag{7.7}$$

where P_0 is the gas permeability of the pure polymer matrix. Different from the highly ordered laminate structure, the orientation of the GO/rGO sheet is normally random without special treatment. The sheet perpendicular to the membrane plane has no "barricade" effect. Thus, the most ideal orientation should be parallel to the membrane plane. To discuss the orientation effect

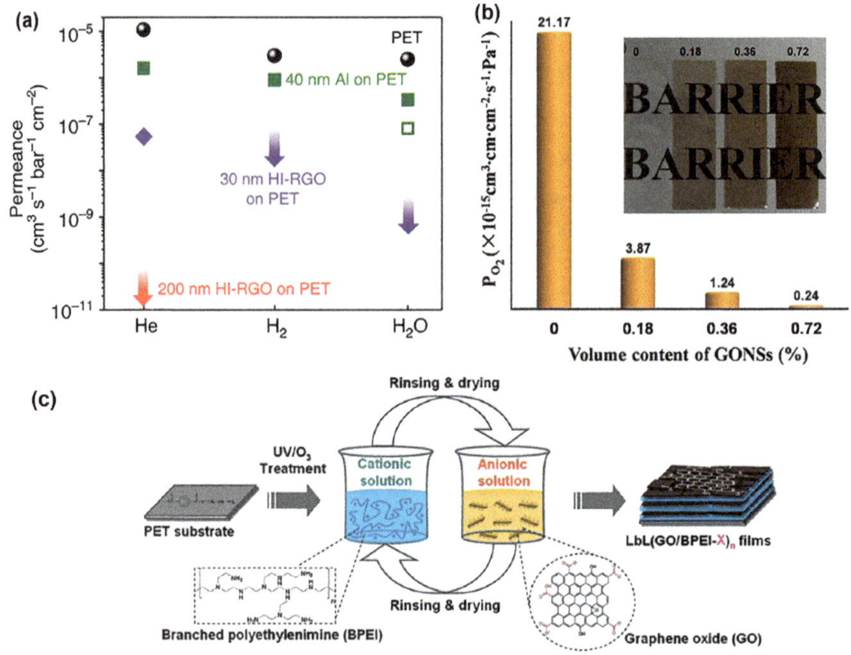

Figure 7.7 (a) Barrier properties of bare PET, HI-RGO on PET and aluminized PET with respect to He, H_2 and H_2O. The solid green symbols are for a 40-nm-thick aluminium film on PET (our measurements) and the open symbol is for 30-nm Al on PET. The violet and red arrows indicate our detection limits for 30 and 200 nm HI-RGO, respectively. Sample-to-sample variations were less than 20%. Reprinted by permission of Macmillan Publishers Ltd: *Nature Communications* (Y. Su, V. Kravets, S. Wong, J. Waters, A. Geim and R. Nair. *Nat. Commn.*, 2014, **5**) Copyright 2014. (b) Permeability of oxygen gas of GO/PVA composites as a function of GO content (insert: semi-transparent GO/PVA film). Reprinted from *Journal of Membrane Science*, **409**, H.-D. Huang, P.-G. Ren, J. Chen, W.-Q. Zhang, X. Ji and Z.-M. Li, High barrier graphene oxide nanosheet/poly(vinyl alcohol) nanocomposite films, 156–163, Copyright 2012, with permission from Elseiver.[51] (c) Schematic of layer-by-layer methods to fabricate GO/PEI barrier. Reprinted from Ref. 52 with permission from the Royal Society of Chemistry.

(Figure 7.6c), an order factor S' is introduced into the description of tortuosity as follows:

$$\tau = 1 + \alpha(S' + 0.5)\varphi/3M \tag{7.8}$$

where $\alpha = L/W$, representing the ratio of the length and thickness of the GO/rGO nanosheet, and M denotes the degree of restacking. If all nanosheets are parallel to the membrane plane, S' equals 1. When the nanosheets all stand perpendicular to the membrane plane, S' is -0.5. If the orientations of the nanosheet are isotropic, $S = 0$.

On the basis of the above theories, the strategies to reduce the permeability (P) are very clear: the first way is to increase the volume fraction of the impermeable nanosheet and to avoid re-aggregation; the second is to increase the value of α by using large GO/rGO sheets; the third is to control the anisotropy of nanosheets in the matrix. By simple melting or mixing, the polymer composites with GO/rGO have already shown significant improvement in their barrier performance (Figure 7.7b).[51,57–59] However, the re-aggregation of nanosheets in the formation of the membrane remains the main negative factor for the barrier properties. Grafting or *in situ* polymerization is one efficient way to avoid the re-aggregation.[60–62] Rigid polymer (*e.g.*, polyimide) linked to rGO can also improve the barrier properties working under a high pressure environment.[63] In the above method, the orientation of the nanosheets is isotropic. To further optimize the barrier performance of the membrane, the ideal way is to employ surface engineering and make the nanosheets aligned. Yang *et al.* used electrostatic layer-by-layer (LbL) self-assembly to fabricate 20 bilayers of polyethylenimine and GO membrane ($(PEI/GO)_{20}$) to successfully reduce the oxygen permeation rate below 2.5×10^{-20} $cm^3\ cm\ cm^{-2}\ s^{-1}\ Pa^{-1}$ in 100% relative humidity, and their oxygen transmission of 98 nm $(PEI/GO)_{20}$ was only 5% of 179 μm PET.[64] The dramatic improvement is attributed to the anisotropy orientation in the LbL membrane and preventing the moisture-induced swelling by the electrostatic interaction. Therefore, the LbL PEI/GO membrane can achieve high permeation with ultrathin thickness (Figure 7.7c).[52] It is believed that other surface engineering methods will further enhance the gas barrier properties of GO/rGO composites.

7.4.2 Anticorrosion and Antifouling Barriers

There has been a long struggle against the threats of corrosion and fouling since the first industrial revolution. Corrosion is a process during which a metal is converted to high-oxidation states, while fouling is an accumulation of undesired species on solid surfaces. As discussed in the CVD graphene section, an ideal anticorrosion barrier should be able to isolate the metal from oxygen, moisture and salt. Owing to the swelling effect, the laminate structure of GO is unable to block moisture or salt.[65] rGO could be a better anticorrosion layer, because the size of interlayer capillary can be decreased down to 0.36 nm and block the permeation of ions.[50] However, only graphene or rGO is not sufficient for long-term anticorrosion, because galvanic corrosion can occur at the carbon/metal interface.[31] A tiny crack can initiate galvanic corrosion at the carbon/metal interface and cause the graphene coating to break off. In addition, if the laminate structures of multilayer rGO sheets were used as a protective layer, the van der Waals interaction between layers would not be adhesive enough to survive mechanical scratching. Therefore, other additives should be employed to build graphene-related anticorrosion layers.

Mixing with polymer matrix is one way to incorporate GO/rGO nanosheets as anticorrosion coatings, enhancing gas and/or ion barrier performances. (3-aminopropyl)triethoxysilane (APTES) was used to modify

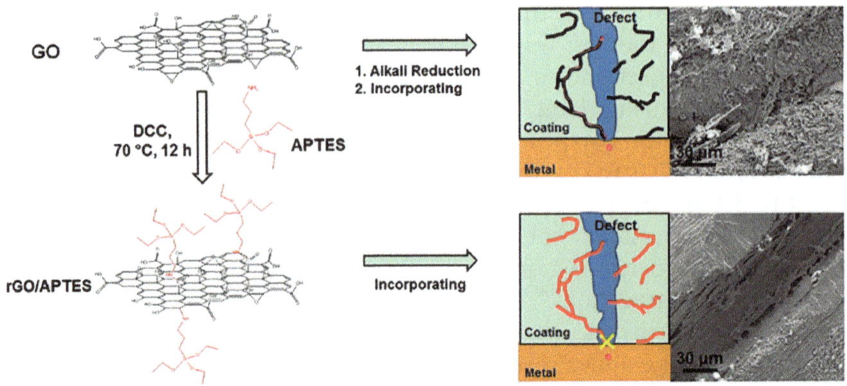

Figure 7.8 Anticorrosion performance of PVB coating combined with rGO and rGO/APTES.
Reprinted with permission from W. Sun, L. Wang, T. Wu, M. Wang, Z. Yang, Y. Pan and G. Liu. *Chem. Mater.*, 2015, **27**(7), 2367, Copyright 2015 American Chemical Society.[66]

rGO as an additive to polyvinylbutyral (PVB) matrix (Figure 7.8).[66] Sun *et al.* found that embedding 5 wt% modified rGO in APTES can reduce the corrosion rate by 65 times with respect to that of the pure PVB coating. An epoxy/rGO composite also acted as an anticorrosion barrier with hydrophobic surfaces.[67,68] With the help of the epoxy/rGO protective layer, the degradation rate can be reduced by two orders. A film of GO/poly(hexamethylene diisocyanate) exhibited robust resistance to corrosion with an inhibition efficiency above 94.3%.[69] A GO paint containing alkyd resin showed a corrosion protection efficiency of about 76% in salt water.[71] Polymeric isocyanate cross-linked with hydroxy functional acrylic adhesive (PIHA) combined with rGO flakes could be electro-deposited on copper and reduced the corrosion rate by more than 9 times.[70] In practical anticorrosion engineering, multilayers of protective barriers are used to perform different functions. It has been shown that GO/rGO or their composites can be very useful when incorporated as gas/ion barriers in anticorrosion coatings as long as direct contact between GO/rGO and the metal surface is avoided to prevent galvanic corrosion until there are further developments.

Fouling issues are widespread in the marine industry, water purification, heat exchangers and other systems. The membrane separation technology for desalination and micro/nano-filtration particularly suffers from fouling problems, because the contaminants will significantly reduce the permeation rate, and even the hydrophobicity/hydrophilicity of the membrane surface can be changed.[71–73] Although a few engineering post-treatments, *e.g.*, back flushing, chemical cleaning, and relaxing, have been developed to recover the membrane, solutions based on fundamental materials are highly desired.[74] Since 2008, use of the tributyltin (TBT) moiety has been banned, and therefore, an alternative antifouling barrier is urgently needed. To avoid organic or inorganic contaminants, the ideal surface should be self-cleaning, amphiphilic and biocidal.

One type of unwanted accumulation is chemicals, *e.g.*, inorganic precipitation or lipids. GO-related coatings have shown excellent resistance to chemical contaminants. GO has been used as an antifouling layer in oil/water separation to avoid the contamination of olein, due to its oleophilic behavior.[75,76] Protected by only 10–15 nm of GO, the permeation of a polyamide ultrafiltration membrane showed no significant degradation (Figure 7.9a).[75] A thin water film on the hydrophilic GO surface acted as a barrier to the infiltration of oil into the polymer matrix, and the ripples on the GO surface contributed nanoscale roughness to make the membrane self-cleaning. Interestingly, the antifouling properties disappeared for the GO protective layer over 80 nm, because the microscale roughness was reduced in the thick laminates. The laminate of rGO also showed excellent resistance to contamination in oil-in-water emulsion separation.[77] Associated with photocatalytic TiO_2, GO membrane can automatically degrade the precipitations to avoid fouling effects in water purification.[78] Depositing a film of APTES/GO nanosheet could improve the antifouling performance of PVDF membrane for the ultrafiltration of proteins.[79] The permeation of PES ultrafiltration membrane embedded with 0.5 wt% GO remained more than 90% after being seriously fouled five times.[80] Similar strategies have been applied to improve the antifouling of other polymer separation membranes.[81–83]

The accumulation of microorganisms is the other origin of fouling issues. A biofouling barrier should either reject the absorption of microorganisms or be biocidal. Graphene has strong cytotoxicity toward bacteria.[84] GO/rGO and their composites also exhibit excellent resistance to biofouling. GO paint has been used to block the colonization of microorganisms.[71] 1 wt% GO in polysulfone can reduce the formation of bio-film (Figure 7.9b).[85] It was proposed that the negative charging on GO makes it more difficult to absorb microorganisms due to electrostatic repulsion. Grafting one layer of GO on polyamide membranes inactivated 60% of bacteria in 1 hour.[86] Considering the above advantages, GO-related materials show great potential in antifouling engineering.

Up to now, various GO-related membranes have been successfully applied in advanced protective coating against gas leakage, corrosion and fouling issues. One future direction is multifunctional barriers using GO-related materials. With a single coating layer that can act as a gas barrier, anticorrosion and antifouling layer at the same time, the coating process would be greatly simplified, and the cost would be significantly reduced. In the future, investigators will be expected to give a full evaluation on the multifunctional performance of GO-related barriers. Additionally, the potential hazardous risks of GO-related barriers should also be evaluated. Overall, the commercialization of GO-related barriers can be foreseen, and more efforts from material scientists, engineers and technicians are required.

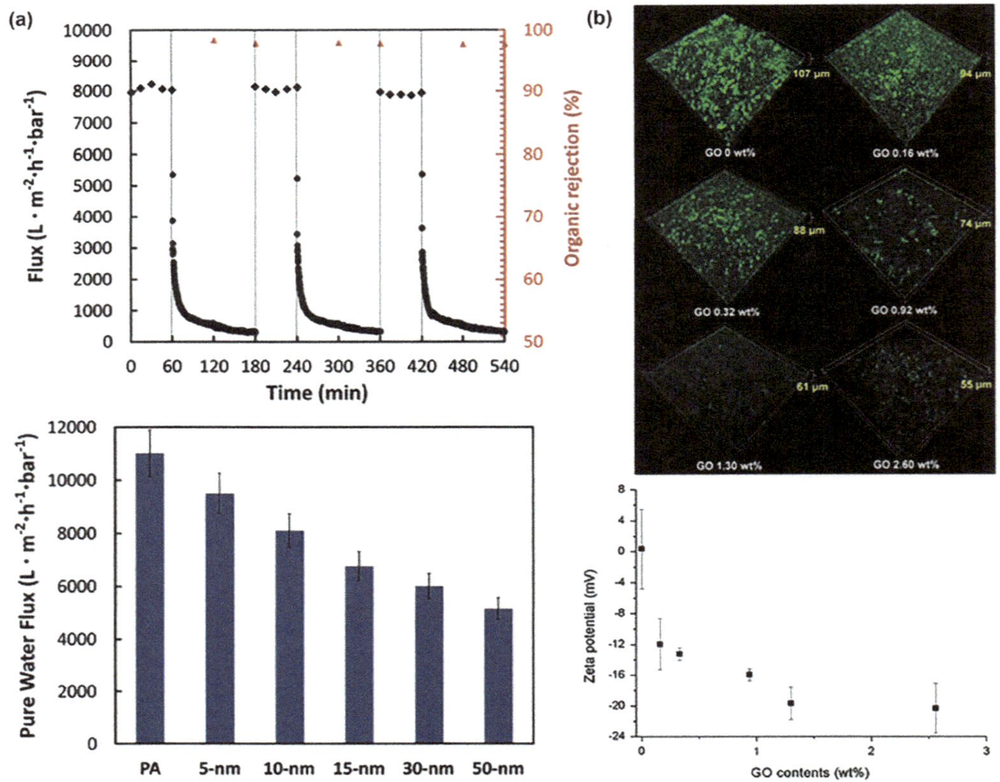

Figure 7.9 (a) The olein resistance and water flux of oil/water separation of polyamide (PA) with 10-nm GO coating, and water permeation rate through the PA with GO coatings of various thicknesses. Reproduced from ref. 84 with permission from John Wiley and Sons, © 2014 WILEY-VCH Verlag GmbH & Co. KGaA, Weinheim. (b) Bacterial resistance of PES membrane containing various GO contents and the zeta potentials of the membranes. Reprinted from *Journal of Membrane Science*, **448**, J. Lee, H.-R. Chae, Y. J. Won, K. Lee, C.-H. Lee, H. H. Lee, I.-C. Kim and J.-M. Lee, Graphene oxide nanoplatelets composite membrane with hydrophilic and antifouling properties for wastewater treatment, 223–230, Copyright 2013, with permission from Elsevier.[85]

7.5 Conclusions and Outlook

The theoretically perfect properties of graphene have been the basis for many studies using graphene as an ideal barrier material. Mechanically exfoliated graphene with a perfect lattice has demonstrated its impermeable nature to all atoms or molecules except for protons. However, the low yield of graphene sheets obtained from the mechanical exfoliation method is not suitable for practical industrial applications. Therefore, mass production of high-quality graphene has become the key challenge for barrier applications.

Although polycrystalline CVD graphene performs well as an anticorrosion barrier for short periods, its intrinsic defects and grain boundaries result in more extensive wet corrosion than uncoated metal surfaces. Apparently, the efforts of synthesizing defect-free single-crystal CVD graphene have been one way to solve this problem. Another choice is to develop new techniques to passivate the defects in graphene sheets to enhance the durable mitigation of corrosion of the CVD graphene barrier.

The application of CVD graphene as a solid-state diffusion barrier for common semiconductor-based conductors in microelectronics has been studied in recent years. For the graphene diffusion barrier tested on Si substrates, an effective barrier can be achieved through the formation of a carbide layer that is chemisorbed on metals. The interfacial properties of the graphene in contact with a diffusing species as well as the bulk substrate are the key parameters in determining whether pristine graphene can prevent diffusion across its interface. A better understanding of the mechanisms that influence graphene's capability of being an effective diffusion barrier in microelectronic applications is required, and continued research is required to improve the adhesion between graphene and the interfaces. Moreover, integration of graphene with other two dimensional materials such as MoS_2 or WSe_2[87,88] would also be beneficial for barrier applications in microelectronics for the purpose of improving the contact properties.

Owing to the achievements in their mass production, GO-related membranes have been successfully applied in advanced protective coatings against gas leakage, corrosion and fouling issues. The corrosion caused by moisture/oxygen can be avoided by coating a laminate of reduced graphene oxide. The addition of GO/rGO significantly improves the gas barrier and anticorrosion performance of polymer coatings, attributed to the "barricade" effects of GO/rGO in the polymer matrix. Chemical and biofouling can be reduced with the assistance of GO/rGO additives. The GO-related membranes exhibit great potential in protective coatings, but cracks on the membrane should be minimized to enhance the stability of the barrier performance. Further engineering works are required to optimize the mechanical strength of GO-related coatings for practical applications. With a deeper understanding of the barrier performance of GO-related membranes, economic and multifunctional protective coatings will be widely employed in the future.

Overall, the prospects for graphene's future in various barrier applications are positive. The industrialization of graphene coatings can be foreseen, especially for GO-related membranes. The future of graphene-based barrier membranes will ultimately advance along with developments in methods and technologies.

Acknowledgements

Support from the National Natural Science Foundation of China (11374244), the Fundamental Research Funds for the Central Universities, the Research Funds of Renmin University of China (17XNLF02), and the Foundation for the Author of National Excellent Doctoral Dissertation of PR China (2014043) is acknowledged.

References

1. T. Sreeprasad and V. Berry, *Small*, 2013, **9**(3), 341.
2. C. Lee, X. Wei, J. W. Kysar and J. Hone, *Science*, 2008, **321**(5887), 385.
3. A. A. Balandin, S. Ghosh, W. Bao, I. Calizo, D. Teweldebrhan, F. Miao and C. N. Lau, *Nano Lett.*, 2008, **8**(3), 902.
4. E. Stolyarova, D. Stolyarov, K. Bolotin, S. Ryu, L. Liu, K. Rim, M. Klima, M. Hybertsen, I. Pogorelsky and I. Pavlishin, *Nano Lett.*, 2008, **9**(1), 332.
5. R. R. Nair, P. Blake, A. N. Grigorenko, K. S. Novoselov, T. J. Booth, T. Stauber, N. M. Peres and A. K. Geim, *Science*, 2008, **320**(5881), 1308.
6. J. S. Bunch, S. S. Verbridge, J. S. Alden, A. M. Van Der Zande, J. M. Parpia, H. G. Craighead and P. L. McEuen, *Nano Lett.*, 2008, **8**(8), 2458.
7. V. Berry, *Carbon*, 2013, **62**, 1.
8. O. Leenaerts, B. Partoens and F. Peeters, *Appl. Phys. Lett.*, 2008, **93**(19), 193107.
9. S. Hu, M. Lozada-Hidalgo, F. Wang, A. Mishchenko, F. Schedin, R. Nair, E. Hill, D. Boukhvalov, M. Katsnelson and R. Dryfe, *Nature*, 2014, **516**(7530), 227.
10. X. Li, W. Cai, J. An, S. Kim, J. Nah, D. Yang, R. Piner, A. Velamakanni, I. Jung and E. Tutuc, *Science*, 2009, **324**(5932), 1312.
11. Q. Yu, J. Lian, S. Siriponglert, H. Li, Y. P. Chen and S.-S. Pei, *Appl. Phys. Lett.*, 2008, **93**, 113103.
12. B. J. Kang, J. H. Mun, C. Y. Hwang and B. J. Cho, *J. Appl. Phys.*, 2009, **106**(10), 104309.
13. S. Chen, W. Cai, R. D. Piner, J. W. Suk, Y. Wu, Y. Ren, J. Kang and R. S. Ruoff, *Nano Lett.*, 2011, **11**(9), 3519.
14. H. Kim, I. Song, C. Park, M. Son, M. Hong, Y. Kim, J. S. Kim, H.-J. Shin, J. Baik and H. C. Choi, *ACS Nano*, 2013, 7(8), 6575.
15. M. Wang, S. K. Jang, W. J. Jang, M. Kim, S. Y. Park, S. W. Kim, S. J. Kahng, J. Y. Choi, R. S. Ruoff and Y. J. Song, *Adv. Mater.*, 2013, **25**(19), 2746.

16. H. J. Song, M. Son, C. Park, H. Lim, M. P. Levendorf, A. W. Tsen, J. Park and H. C. Choi, *Nanoscale*, 2012, **4**(10), 3050.
17. D. L. Duong, G. H. Han, S. M. Lee, F. Gunes, E. S. Kim, S. T. Kim, H. Kim, Q. H. Ta, K. P. So and S. J. Yoon, *Nature*, 2012, **490**(7419), 235.
18. S. P. Koenig, L. Wang, J. Pellegrino and J. S. Bunch, *Nat. Nanotechnol.*, 2012, 7(11), 728.
19. S. C. O'Hern, C. A. Stewart, M. S. Boutilier, J.-C. Idrobo, S. Bhaviripudi, S. K. Das, J. Kong, T. Laoui, M. Atieh and R. Karnik, *ACS Nano*, 2012, **6**(11), 10130.
20. D. Berman, A. Erdemir and A. V. Sumant, *Carbon*, 2013, **54**, 454.
21. D. Berman, S. A. Deshmukh, S. K. Sankaranarayanan, A. Erdemir and A. V. Sumant, *Adv. Funct. Mater.*, 2014, **24**(42), 6640.
22. W. Zhang, S. Lee, K. L. McNear, T. F. Chung, S. Lee, K. Lee, S. A. Crist, T. L. Ratliff, Z. Zhong and Y. P. Chen, *Sci.Rep.*, 2014, **4**, 4097.
23. J. Gray and B. Luan, *J. Alloys Compd.*, 2002, **336**(1), 88.
24. M. Redondo and C. B. Breslin, *Corros. Sci.*, 2007, **49**(4), 1765.
25. V. Mittal, S. Bera, T. Saravanan, S. Sumathi, R. Krishnan, S. Rangarajan, S. Velmurugan and S. Narasimhan, *Thin Solid Films*, 2009, **517**(5), 1672.
26. P. J. Kinlen, V. Menon and Y. Ding, *J. Electrochem. Soc.*, 1999, **146**(10), 3690.
27. G. Grundmeier, C. Reinartz, M. Rohwerder and M. Stratmann, *Electrochim. Acta*, 1998, **43**(1), 165.
28. M. Segarra, L. Miralles, J. Diaz, H. Xuriguera, J. Chimenos, F. Espiell and S. Pinol, *Mater. Sci. Forum*, 2003, **426**, 3511.
29. S. Chen, L. Brown, M. Levendorf, W. Cai, S. Y. Ju, J. Edgeworth, X. Li, C. W. Magnuson, A. Velamakanni and R. D. Piner, *ACS Nano*, 2010, **5**(2), 1321.
30. D. Prasai, J. C. Tuberquia, R. R. Harl, G. K. Jennings and K. I. Bolotin, *ACS Nano*, 2012, **6**(2), 1102.
31. M. Schriver, W. Regan, W. J. Gannett, A. M. Zaniewski, M. F. Crommie and A. Zettl, *ACS Nano*, 2013, 7(7), 5763.
32. Y.-P. Hsieh, M. Hofmann, K.-W. Chang, J. G. Jhu, Y.-Y. Li, K. Y. Chen, C. C. Yang, W.-S. Chang and L.-C. Chen, *ACS Nano*, 2014, **8**(1), 443.
33. F. Zhou, Z. Li, G. J. Shenoy, L. Li and H. Liu, *ACS Nano*, 2013, 7(8), 6939.
34. H.-Y. Kim, C. Lee, J. Kim, F. Ren and S. Pearton, *J. Vac. Sci. Technol., B: Nanotechnol. Microelectron.: Mater., Process., Meas., Phenom.*, 2012, **30**(3), 030602.
35. J. Eun Lee, B. K. Sharma, S.-K. Lee, H. Jeon, B. Hee Hong, H.-J. Lee and J.-H. Ahn, *Appl. Phys. Lett.*, 2013, **102**(11), 113112.
36. C. P. Y. Wong, T. J. H. Koek, Y. Liu, K. P. Loh, K. E. J. Goh, C. Troadec and C. A. Nijhuis, *ACS. Appl. Mater. Interfaces*, 2014, **6**(22), 20464.
37. C. G. Kang, S. K. Lim, S. Lee, S. K. Lee, C. Cho, Y. G. Lee, H. J. Hwang, Y. Kim, H. J. Choi and S. H. Choe, *Nanotehnology*, 2013, **24**(11), 115707.

38. J. Hong, S. Lee, S. Lee, H. Han, C. Mahata, H.-W. Yeon, B. Koo, S.-I. Kim, T. Nam and K. Byun, *Nanoscale*, 2014, **6**(13), 7503.
39. Y. Zhao, Z. Liu, T. Sun, L. Zhang, W. Jie, X. Wang, Y. Xie, Y. H. Tsang, H. Long and Y. Chai, *ACS Nano*, 2014, **8**(12), 12601.
40. B.-S. Nguyen, J.-F. Lin and D.-C. Perng, *Appl. Phys. Lett.*, 2014, **104**(8), 082105.
41. J. H. Bong, S. J. Yoon, A. Yoon, W. S. Hwang and B. J. Cho, *Appl. Phys. Lett.*, 2015, **106**(6), 063112.
42. D. A. Dikin, S. Stankovich, E. J. Zimney, R. D. Piner, G. H. Dommett, G. Evmenenko, S. T. Nguyen and R. S. Ruoff, *Nature*, 2007, **448**(7152), 457.
43. D. R. Dreyer, S. Park, C. W. Bielawski and R. S. Ruoff, *Chem. Soc. Rev.*, 2010, **39**(1), 228.
44. J. Chong, B. Wang, C. Mattevi and K. Li, *J. Membr. Sci.*, 2018, **549**, 385.
45. G. Eda, G. Fanchini and M. Chhowalla, *Nat. Nanotechnol.*, 2008, **3**(5), 270.
46. J. T. Robinson, M. Zalalutdinov, J. W. Baldwin, E. S. Snow, Z. Wei, P. Sheehan and B. H. Houston, *Nano Lett.*, 2008, **8**(10), 3441.
47. R. Nair, H. Wu, P. Jayaram, I. Grigorieva and A. Geim, *Science*, 2012, **335**(6067), 442.
48. H. W. Kim, H. W. Yoon, S.-M. Yoon, B. M. Yoo, B. K. Ahn, Y. H. Cho, H. J. Shin, H. Yang, U. Paik and S. Kwon, *Science*, 2013, **342**(6154), 91.
49. H. Li, Z. Song, X. Zhang, Y. Huang, S. Li, Y. Mao, H. J. Ploehn, Y. Bao and M. Yu, *Science*, 2013, **342**(6154), 95.
50. Y. Su, V. Kravets, S. Wong, J. Waters, A. Geim and R. Nair, *Nat. Commun.*, 2014, **5**, 4843.
51. H.-D. Huang, P.-G. Ren, J. Chen, W.-Q. Zhang, X. Ji and Z.-M. Li, *J. Membr. Sci.*, 2012, **409**, 156.
52. J.-T. Chen, Y.-J. Fu, Q.-F. An, S.-C. Lo, S.-H. Huang, W.-S. Hung, C.-C. Hu, K.-R. Lee and J.-Y. Lai, *Nanoscale*, 2013, **5**(19), 9081.
53. W.-S. Hung, C.-H. Tsou, M. De Guzman, Q.-F. An, Y.-L. Liu, Y.-M. Zhang, C.-C. Hu, K.-R. Lee and J.-Y. Lai, *Chem. Mater.*, 2014, **26**(9), 2983.
54. H. W. Kim, H. W. Yoon, B. M. Yoo, J. S. Park, K. L. Gleason, B. D. Freeman and H. B. Park, *Chem. Commun.*, 2014, **50**(88), 13563.
55. Y. Cui, S. Kundalwal and S. Kumar, *Carbon*, 2016, **98**, 313.
56. B. M. Yoo, H. J. Shin, H. W. Yoon and H. B. Park, *J. Appl. Polym. Sci.*, 2014, **131**(1), 39628.
57. H. Kim, Y. Miura and C. W. Macosko, *Chem. Mater.*, 2010, **22**(11), 3441.
58. P. a. Song, Y. Yu, T. Zhang, S. Fu, Z. Fang and Q. Wu, *Ind. Eng. Chem. Res.*, 2012, **51**(21), 7255.
59. H. Huang, P. Ren, J. Xu, L. Xu, G. Zhong, B. Hsiao and Z. Li, *J. Membr. Sci.*, 2014, **464**, 110.
60. O. C. Compton, S. Kim, C. Pierre, J. M. Torkelson and S. T. Nguyen, *Adv. Mater.*, 2010, **22**(42), 4759.

61. X. Wang, Y. Hu, L. Song, H. Yang, W. Xing and H. Lu, *J. Mater. Chem.*, 2011, **21**(12), 4222.
62. C. Xiang, P. J. Cox, A. Kukovecz, B. Genorio, D. P. Hashim, Z. Yan, Z. Peng, C.-C. Hwang, G. Ruan and E. L. Samuel, *ACS Nano*, 2013, 7(11), 10380.
63. J. Lim, H. Yeo, M. Goh, B.-C. Ku, S. G. Kim, H. S. Lee, B. Park and N.-H. You, *Chem. Mater.*, 2015, **27**(6), 2040.
64. Y. H. Yang, L. Bolling, M. A. Priolo and J. C. Grunlan, *Adv. Mater.*, 2013, **25**(4), 503.
65. R. Joshi, P. Carbone, F.-C. Wang, V. G. Kravets, Y. Su, I. V. Grigorieva, H. Wu, A. K. Geim and R. R. Nair, *Science*, 2014, **343**(6172), 752.
66. W. Sun, L. Wang, T. Wu, M. Wang, Z. Yang, Y. Pan and G. Liu, *Chem. Mater.*, 2015, **27**(7), 2367.
67. K.-C. Chang, M.-H. Hsu, H.-I. Lu, M.-C. Lai, P.-J. Liu, C.-H. Hsu, W.-F. Ji, T.-L. Chuang, Y. Wei and J.-M. Yeh, *Carbon*, 2014, **66**, 144.
68. L. Gu, S. Liu, H. Zhao and H. Yu. *ACS. Appl. Mater. Interfaces*, 2015, 7 (32), 17641.
69. S. C. Sahu, A. K. Samantara, M. Seth, S. Parwaiz, B. P. Singh, P. C. Rath and B. K. Jena, *Electrochem. Commun.*, 2013, **32**, 22.
70. B. P. Singh, S. Nayak, K. K. Nanda, B. K. Jena, S. Bhattacharjee and L. Besra, *Carbon*, 2013, **61**, 47.
71. K. Krishnamoorthy, K. Jeyasubramanian, M. Premanathan, G. Subbiah, H. S. Shin and S. J. Kim, *Carbon*, 2014, **72**, 328.
72. W. Gao, H. Liang, J. Ma, M. Han, Z.-L. Chen, Z.-S. Han and G.-B. Li, *Desalination*, 2011, **272**(1), 1.
73. P. Xu, J. E. Drewes, T.-U. Kim, C. Bellona and G. Amy, *J. Membr. Sci.*, 2006, **279**(1), 165.
74. T. Zsirai, P. Buzatu, P. Aerts and S. Judd, *Water Res.*, 2012, **46**(14), 4499.
75. Y. Huang, H. Li, L. Wang, Y. Qiao, C. Tang, C. Jung, Y. Yoon, S. Li and M. Yu, *Adv. Mat. Inter.*, 2015, **2**, 1400433.
76. Y. Dong, J. Li, L. Shi, X. Wang, Z. Guo and W. Liu, *Chem. Commun.*, 2014, **50**(42), 5586.
77. N. Liu, M. Zhang, W. Zhang, Y. Cao, Y. Chen, X. Lin, L. Xu, C. Li, L. Feng and Y. Wei, *J. Mater. Chem. A.*, 2015, **3**(40), 20113.
78. C. Xu, Y. Xu and J. Zhu, *ACS. Appl. Mater. Interfaces.*, 2014, **6**(18), 16117.
79. Z. Xu, J. Zhang, M. Shan, Y. Li, B. Li, J. Niu, B. Zhou and X. Qian, *J. Membr. Sci.*, 2014, **458**, 1.
80. S. Zinadini, A. A. Zinatizadeh, M. Rahimi, V. Vatanpour and H. Zangeneh, *J. Membr. Sci.*, 2014, **453**, 292.
81. J. Zhang, Z. Xu, M. Shan, B. Zhou, Y. Li, B. Li, J. Niu and X. Qian, *J. Membr. Sci.*, 2013, **448**, 81.
82. H. Zhao, L. Wu, Z. Zhou, L. Zhang and H. Chen, *Phys. Chem. Chem. Phys*, 2013, **15**(23), 9084.
83. S. Bano, A. Mahmood, S.-J. Kim and K.-H. Lee, *J. Mater. Chem. A.*, 2015, **3**(5), 2065.

84. S. Liu, T. H. Zeng, M. Hofmann, E. Burcombe, J. Wei, R. Jiang, J. Kong and Y. Chen, *ACS Nano*, 2011, **5**(9), 6971.
85. J. Lee, H.-R. Chae, Y. J. Won, K. Lee, C.-H. Lee, H. H. Lee, I.-C. Kim and J.-M. Lee, *J. Membr. Sci.*, 2013, **448**, 223.
86. F. Perreault, M. E. Tousley and M. Elimelech, *Environ. Sci. Technol. Lett.*, 2013, **1**(1), 71.
87. G. Fiori, F. Bonaccorso, G. Iannaccone, T. Palacios, D. Neumaier, A. Seabaugh, S. K. Banerjee and L. Colombo, *Nat. Nanotechnol.*, 2014, **9**(10), 768.
88. S. Das, J. A. Robinson, M. Dubey, H. Terrones and M. Terrones, *Annu. Rev. Mater.*, 2015, **45**, 1.

CHAPTER 8

Graphene-based Membranes for Purification and Separation Applications

XIAOHENG JIN, CHUNG-YAN POON AND RAKESH K. JOSHI*

SMaRT Center, School of Materials Science and Engineering, University of New South Wales, Sydney, NSW 2052, Australia
*Email: r.joshi@unsw.edu.au

8.1 Introduction

Mass transport through a porous structure has extensive applications in areas like water purification, toxic gas absorption, and desalination.[1] Membranes based on materials such as polymers and inorganics, and metal organic frameworks have been extensively utilized in the past decade for mass transport applications.[2–9] Those varieties of material-based membranes can share similar macro-structures but very different micro-structures.[10] For example, zeolite membranes have a high gas permeability and structural integrity with a thickness of 50 μm[11] that can also be controlled to around 1 μm. Their pore size can be controlled to 8–12 atom rings by growth of the zeolite on a porous support with the help of a structure-directing agent (SDA).[12] However, to achieve high-quality zeolite membranes, the fabrication can be very expensive.

Polymer membranes, for example, polycarbonate and polyester, were first applied in gas separation thanks to their workable characteristics and controllable gas permeability by designing the polymer molecule.[9] However, most of the polymers cannot survive high temperatures and harsh chemical

Nanoscience & Nanotechnology Series No. 47
Graphene-based Membranes for Mass Transport Applications
Edited by Pengzhan Sun and Hongwei Zhu

Published by the Royal Society of Chemistry, www.rsc.org

environments; some even suffer plasticization, which prevents polymer membranes from being efficiently used in industrial applications.[9] Moreover, polymer membranes naturally yield a permeability and selectivity below the upper bounds discovered by Robeson in 1991 (updated in 2008).[13] Freeman and co-workers then explained the upper bound as a consequence of the sieving nature of the glass-like polymer chain.[14]

Another exemplary polymer for mass transport applications is polysulfone (PSf).[15] This easy-to-manufacture, high thermally stable, and chemically inert material can act both as the matrix of a mass transport membrane and a filler of it.[9] Polysulfone is currently used in the air separation industries, natural gas purification, chemical absorption processes and so on, despite the problem of biofouling, which is the nature of accumulation of microorganism and proteins.[15] Those side effects might block the solution from the membranes, hence raising the maintenance costs while lowering the efficiency of mass transport.

Restrained by their micro-structure and 2-dimensional shape, membranes, thin films, and nanosheets are mostly applied to gas separation and water desalination.[16] Hybrid organic–inorganic membranes allow fuel cell to work at a higher efficiency; polymeric membranes have been applied in ultrafiltration, nanofiltration, reverse osmosis and gas separation; CO_2 capture/removal and H_2 generation also need inorganic membranes, *etc.*[8,17,18]

Mass transport membranes also enable advances in green industries like seawater desalination, natural gas separation, waste product filtration and so on.[19–22] Besides the extraction of resources like gas and water from natural mixtures, membranes are used in industrial processes to reduce their energy consumption as well. For instance, in a sequence of organic chemistry reactions, solvents can be different in continuous steps, and the traditional way to exchange them is boiling the low-boiling-point solvent of the former step, then adding the high-boiling-point solvent of the later step. If the boiling point of the two solvents is reversed, energy consumption would be steeply raised to keep the same procedure. However, membranes can allow organic solvent nanofiltration to exchange these two solvents in a fast and conservative way, and no solvent needs to be boiled as it is filtrated at room temperature.[23]

8.2 Graphene-based Membranes

Graphite is a 3D material formed by stacking multiple layers of graphene in an AB sequence where each layer is connected by a weak van der Waals force.[24–26] Therefore, in graphite, the graphene basal planes can shear or glide over each other, giving it the characteristic of lubrication. Graphene is a one-atom-thick layer of bonded sp2 (trigonal) carbon atoms packed into a two-dimensional honeycomb lattice.[27] As graphene is a very highly ordered material, it has few defects on its surface.[28] According to a review from

Yan *et al.*,[29] graphene has unique physical properties, such as a tunable band gap, quantum Hall effect, and electron confinement effects.[29] Furthermore, it shows outstanding performance in mechanical strength, electrical and thermal conductivity, and optical transparency.[25] With a characteristic of delocalized electrons, graphene has a high electrical conductivity. The value could be higher by improving the impurity scattering. The experimental Young modulus is 1 TPa, and the strength is 130 GPa.[28] These special features of graphene make it very attractive. Comparing to CNTs, graphene has a much higher chemical inertness. Thus, the reactivity of graphene to other chemicals is very low.[29]

There are two conventional approaches to produce graphene: the bottom-up and top-down. In the former approach, graphene sheets are built from assembled carbon atoms. The top-down approach comprises extracting a graphene layer from its 3D form graphite.[30]

Graphene can be considered as an ideal membrane since the thickness is only one carbon diameter, which is theoretically expected to give the highest water flux. However, pristine defect-free monolayer graphene is impermeable to gaseous and liquid molecules. Thus, one of the possible ways to generate a water path is by creating nanopores in the graphene sheets. Such nanopores can be produced by irradiation with an electron beam,[31] diblock copolymer templating, ultraviolet-induced oxidative etching, helium ion beam drilling, and chemical etching.[27,32,33] It is estimated that the water permeability of graphene membrane is 2–3 orders of magnitude greater than commercial reverse osmosis membranes.[34] After nanopores are created, the inherent hydrophobia of the parent graphene can create a capillary force for water permeation. Recently, it has been shown that nanoporous graphene can be used as a reverse osmosis membrane regarding both mechanical and separation performances, such as salt rejection and water flux. The water can flow across a nanoporous graphene membrane at rates in the range of 10–100 $L\,cm^{-2}\,day^{-1}\,MPa^{-1}$ with a rejection rate 2 to 3 orders of magnitude higher than diffusive RO membranes.[34] Also, chemical functionalization of graphene nanopores, such as nitrogen and fluorine decorated graphene, could selectively reject specific solvated ions due to the attraction and repulsion of the ion charges.[35] Graphene oxide, carbon nanotubes, reduced graphene oxide and chemically converted graphene are also examples of functionalized graphene.[36–38]

8.2.1 Challenges and Opportunities for Graphene Family Membranes

The discovery of graphene in 2004 launched an exploration of its excellent properties.[39] However, pure graphene is almost completely impermeable to liquids and gases, because the carbon atoms are tightly packed in the graphene lattice, and they are prohibited from interactions with other atoms by the electron cloud on both of its sides.[40]

Researchers have attempted to make water flow through the interlayer of graphene membranes, but most of them worked on regular oxidation of graphene membranes to achieve a high-speed water flow and expanded interlayer galleries.[39–41] Graphene oxide and reduced graphene oxide membranes have been introduced to the mass transport field over the past several years; however, designing high-permeability, high-selectivity, mechanically stable and renewable membranes is still a challenge.[19,42]

8.3 Graphene Oxide Membranes

Graphene oxide membrane, with an abundance of functionalized groups and vacancies attached on or within the sheets, has tuneable microstructures and physicochemical properties that offer extraordinary potential in the fields of energy, environment, and water-related applications. Moreover, it is possible to tailor the microstructure by manipulating the interlayer spacing and controlling the oxidized area of GO during synthesis to meet the demand of the current devices or the future designs. It should be noted that graphene oxide is a hydrophilic oxygenated single layer material, whereas graphite oxide refers to stacked graphene oxide flakes.

One of the most important applications of graphene oxide is membrane filtration. Unlike graphene, which is impermeable to all gases and molecules, graphene oxide has a certain extent of permeability and permselectivity depending on the oxidation level, interlayer spacing, and the water content. The oxidized region, which refers to an area with high hydrophilicity, allows GO to be well dispersed in solution as single-layer sheets.[26,43] Additionally, the hydroxyl and epoxy groups create a spacing between the membranes that forms nanocapillaries[42] and the negatively charged polar functional groups attract water or certain molecules,[37] hence, facilitating the permeation rate. On the other hand, the hydrophobic (non-oxidized) part was also found to govern the water flow,[39] due to its frictionless nature. Interlayer spacing is an important parameter in determining the application of water desalination, water purification, molecular sieving, and bacterial inactivation, as schematically illustrated in Figure 8.1. Intriguingly, an enlarged GO spacing can be easily achieved by introducing rigid chemical groups or polymer chains in between the sheets. The presence of water content within the sheets might expand the interlayer spacing.[42] Meanwhile, it can show auxetic material behaviour.[44,45] Nair *et al.* suggested that the nanocapillaries in the GO film become narrower in low water content conditions.[46] These characteristics indicate the possibility of controlling the permeability and permselectivity of GO membrane.

8.3.1 Structure of GO

Graphene oxide is composed of carbon, oxygen, and hydrogen. More specifically, epoxide and hydroxyl groups are attached to the carbon basal plane, while the carbonyl, carboxyl, and phenol groups are found on the edges of

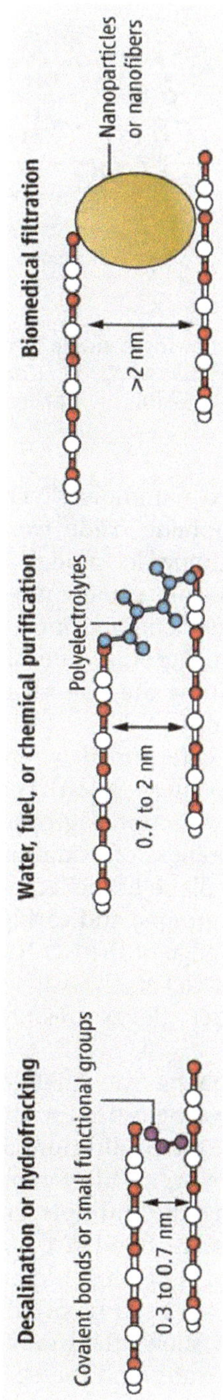

Figure 8.1 The interlayer spacing/nanochannel size can be manipulated for specific applications.[37] Reproduced from ref. 37, reprinted with permission from AAAS.

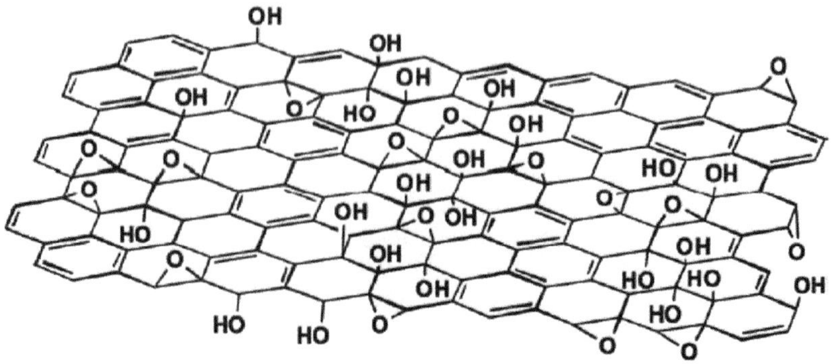

Figure 8.2 Lerf-Klinowski model showing a single layer of graphene oxide.[47] Reproduced from *The Chemistry of Graphene Oxide*, 2015, X. Jin, C.-Y. Poon and R. K. Joshi, © Springer International Publishing Switzerland 2015.

the GO sheets or the embedded nanoholes.[25] Over the past few decades, different lattice models of graphene oxide were suggested. The widely-accepted model is the Lerf-Klinowski model, as shown in Figure 8.2 below.[47] These functional groups are usually negatively charged, which enhances the rejection of negatively charged ions in the application of molecular separation. Also, functional groups on both the edge and basal plane play important roles in determining the water transport paths for filtration and molecular sieving applications.

The structure of GO can be determined by characterization techniques such as nuclear magnetic resonance (NMR) spectroscopy and X-ray diffraction (XRD) analysis.[48] The functional group density is based on the graphite source and the completeness of oxidation that occurred. Graphene oxide is graphene with its basal plane branched to three kinds of functional groups: epoxy groups, hydroxyl groups, and carboxyl groups.[47] The carboxyl groups are only attached to the edge of the GO basal plane. As the branches occupy space, multiple layers of GO stack with spacing between the flakes. Thus, overlapping of the GO flakes results in the formation of nanocapillaries.

The thickness of a single graphene oxide platelet is 3.4 Å.[39] In the form of multiple layers, the interlayer spacing is 9 ± 1 Å without water and 13 ± 1 Å in water.[39] It was also reported that the equilibrium distance of graphene layers in GO is 8.43–8.59 Å separated by an ice bilayer and 11.38–11.46 Å separated by an ice trilayer.[44] Considering the multiple graphene oxide layers, Lerf *et al.* suggested that the layers are bonded by strong hydrogen bonds to maintain the stacked structure. A strong interaction between the oxygen functionality on GO and water has been reported and was characterized by neutron scattering.[47] Figure 8.3 shows the model from Dreyer *et al.*[47] proposing the interaction between water and the epoxide group.[49]

Also, Bouhvalov *et al.*[44] suggested that the ultrafast permeation occurred only when the GO membrane possesses its optimal interlayer distance,

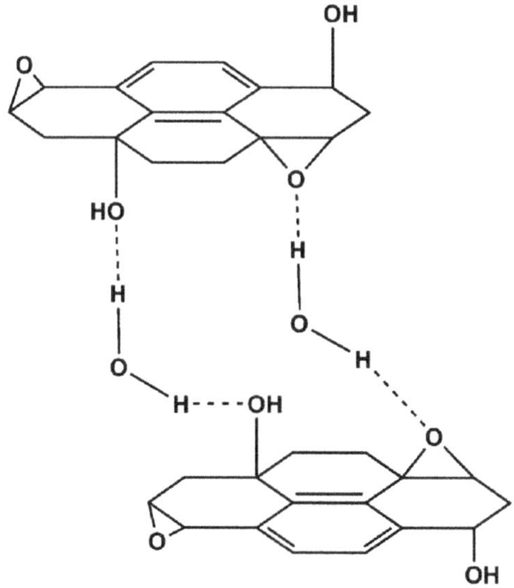

Figure 8.3 The proposed interaction of water and GO. Hydrogen bonds are formed between epoxy groups and water.[47]
Reproduced from *The Chemistry of Graphene Oxide*, 2015, X. Jin, C.-Y. Poon and R. K. Joshi, © Springer International Publishing Switzerland 2015.

which is around 7–10 Å.[44] This is because the ice bilayer is formed in the interlayer space and anisotropic water flowing is allowed. On the other hand, Nair *et al.*[40] suggested a model of GO permeation. The GO laminates form nanocapillaries that are filled with an ordered monolayer of water molecules. Water passes through the graphene sheets only if the interlayer spacing is larger 6 Å. The study of Nair *et al.*[40] found that the interlayer thickness of GO membranes decreases from 10 Å to 4 Å after annealing at 300 °C in a hydrogen–argon atmosphere.[40] This indicates that the presence of water molecules in GO has expanded the GO films.

8.3.2 Preparation of Graphene Oxide

The structure of the graphene oxide sheet is still a controversy today. One of the reasons is that the microstructures and properties of graphene oxide are strongly affected by the synthesis methods and fabrication conditions.[50] It can be synthesized by the Staudenmaier, Brodie, Hofmann or Hummers methods,[36,50,51] as summarized below in Table 8.1.[47,50] The most popular method is the Hummers method since graphene oxide can be produced within 2 hours with a high degree of oxidation.

The reactions listed above have different reagents and working conditions, but most of them involve strong oxidizing agents. For example, either concentrated sulphuric acid, concentrated nitric acid, or both are applied to oxidize graphite into graphite oxide.[47,50] Graphene oxide was first prepared

Table 8.1 Summary of different oxidation methods of graphite.

Methods	Year of publication	Chemicals used			
Brodie method	1859	Potassium chlorate	Graphite	Fuming nitric acid	
Staudenmaier method	1898	Potassium chlorate	Graphite	Concentrated sulphuric acid	Fuming nitric acid
Hofmann method	1937	Potassium chlorate	Graphite	Concentrated sulphuric acid	Concentrated nitric acid
Hummers method	1958	Sodium nitrate	Graphite	Concentrated sulphuric acid	Potassium permanganate

by Brodie in 1959. The product shows a light-yellow color, but it is claimed that the product could not be synthesized by one continued treatment. It was then modified and improved by Staudenmaier, where the concentrated sulphuric acid was introduced, and multiple aliquots of potassium chlorate solution were added to the reaction mixture. However, the Staudenmaier method was time-consuming and hazardous. Then, the Hummers method, using a water-free mixture of concentrated sulphuric acid, sodium nitrate, and potassium permanganate, was introduced. However, the product was found to incompletely oxidize. Therefore, pre-expansion or pre-treatment, such as with H_2SO_4, $K_2S_2O_8$, and P_2O_5, is helpful to achieve a higher degree of oxidation. It has also been reported that modifications can be made by increasing the amount of potassium permanganate or by choosing a smaller flake size of graphite, *etc.*[52,53]

As mentioned above, the number of functional groups and the degree of oxidation are varied by the processing methods and conditions.[50] These characteristics determine the interlayer spacing, nature of the nanocapillary, and pore size of the membranes, which then further influence the permeability outcome of the graphene oxide membrane. Table 8.2 shows a comparison of properties between graphite oxide and thermally reduced graphene. Both materials are produced using various methods from the experimental results of Poh *et al.*[50] From the experimental results, graphite oxide prepared with the Hummers method has the greatest extent of oxidation and the largest interlayer spacing.[50,54] This shows that graphite oxide obtained by using the Hummers method can produce graphite oxide with more nanocapillaries and can increase the permeation rate.

The thermally reduced graphene is a single layer of graphene that has been thermally exfoliated inside a vacuum tight tube furnace at 1000 °C.[50] During the experimental process, the graphene is reduced, which leads to a decrease in the degree of oxidation. The microstructure of the reduced graphene is dissimilar to graphene oxide, which has a high C/O ratio. However, the crystallite size of the reduced graphene obtained by Raman spectroscopy should be noticed since it indicates the density of defeats in the graphite oxide. The theory behind this is further discussed in the section on analysis techniques.

Table 8.2 Comparison of graphite oxide and thermally reduced graphene prepared by Staudenmaier, Hofmann and Hummers methods.[50] Reproduced from ref. 50 with permission from the Royal Society of Chemistry.

	Graphite oxide		Thermally reduced graphene	
Methods	C/O ratio	Interlayer spacing (nm)	Crystallite size (nm)	C/O ratio
Staudenmaier	2.52	0.7084	21.8	24.1
Hofmann	1.77	0.7226	15.4	18.22
Hummers	1.12	0.8133	17.3	19.27

There are multiple ways to exfoliate graphite oxide to graphene oxide. It can be done either thermally or mechanically.[36,47] Sonicating is a relatively popular method since the process is simple and requires only a short amount of time. Graphene oxide can be dispersed in polar solvents. The dispersibility is proportional to the polarity of the surface of the graphene oxide sheets, which depends on the extent of the oxidation.[47] Li *et al.*[26] suggested that the highly negative charged GO flakes in aqueous solution repel each other due to electrostatic repulsion, forming stable single layered GO in the form of a colloid.[26] More specifically, hydroxyl groups on the basal plane of the GO sheets play a key role in stabilizing and promoting the dispersibility of the GO solution (Prerna Bansal, *The Journal of Physical Chemistry C*, 2017, **121**, 9847–9859). The presence of oxygen functionalities also further assists the formation of uniform and intact membranes.

There are several conventional methods to fabricate GO membranes.[55] Vacuum filtration appears attractive due to its simplicity. In general, vacuum filtration is used to separate aqueous volatile solutions.[56] For membrane fabrication, a vacuum pump is used to extract the water from the GO solution with a pulling force in addition to the gravitational force. A uniform GO membrane is then formed on the filter paper as a supporting substrate. Dikin *et al.*[57] proposed a mechanism of formation of graphene oxide film by vacuum filtration. At first, the graphene oxide flakes in the colloid are forced onto the surface of the substrate membrane by the water flow.[57] The graphene oxide sheets assemble to form multiple layers of graphene oxide. The layers become more packed during the later stage of filtration.[57] As more water is removed from the suspension, the concentration of graphene oxide increases. This leads to an increase in 'sheet-to-sheet' interactions.[57] More packed and arranged layers of graphene oxide are formed and align on top of each other. The sheets are arranged evenly on the layers by the water flow.[57]

The other methods are layer-by-layer assembly, drop-casting, and spin-coating.[55] In a layer-by-layer assembly, the GO films are stabilized by covalent bonds, electrostatic forces or both.[37] The layer-by-layer assembly requires a cycle of assembly to stack GO layers on the substrate. In Hu and Mi's[58] work, positively charged polyallylamine hydrochloride (PAH) was employed as a polyelectrolyte to bind the GO sheets on the hydrolyzed peroxyacyl nitrate (PAN) substrate by electrostatic interaction, hydrophobic interaction and hydrogen bonding. The substrate was immersed in PAH solution for 30 minutes and then immersed in GO solution for 30 minutes again. This process was repeated several times to form a certain thickness of GO membrane.

8.3.3 Commonly Used Characterization Methods

8.3.3.1 Scanning Electron Microscopy (SEM)

In SEM, the electron beam scans over the sample surface and interacts with the surface and subsurface forming secondary electrons and back-scattered electrons. The electrons are collected and analyzed to give configuration information. The SEM images reveal details of the surface morphology of the membranes, from which the size and density of the pores can be estimated. For instance, the SEM images from Boguslawski *et al.*[59] in Figure 8.4(a) show the uneven surface of a GO membrane and Figure 8.4(b) shows the cross-section of GO layers.[59] The SEM images from Feng *et al.*[60] in Figure 8.5(b) show a cross-section of GO. Layers of GO forming nanocapillaries can be observed. The size of the nanocapillaries is smaller than 1 μm.

8.3.3.2 Atomic Force Microscopy (AFM)

AFM is used to investigate the surface morphology of graphene and graphene oxide. Similar to SEM, the size of the flakes, density of the pores and the shape of the flakes can be investigated.[61] Furthermore, the number of layers can be determined by AFM. For example, the change in thickness before and after graphene oxide reduction was investigated by Jung *et al.*[62] The study showed that the thickness of the multi-layered graphene oxide treated by thermal reduction increases with the number of layers. One possible explanation is that the thicker the layer, the more water molecules

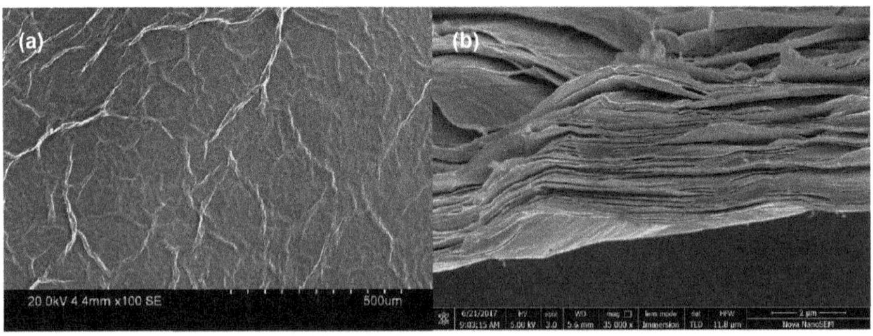

Figure 8.4 SEM images of (a) GO surface and a (b) cross section of GO.[59] Reprinted from *Carbon*, **129**, Y. You, X. H. Jin, X. Y. Wen, V. Sahajwalla, V. Chen, H. Bustamante and R. K. Joshi, Application of graphene oxide membranes for removal of natural organic matter from water, 415–419, Copyright 2018, with permission from Elsevier.

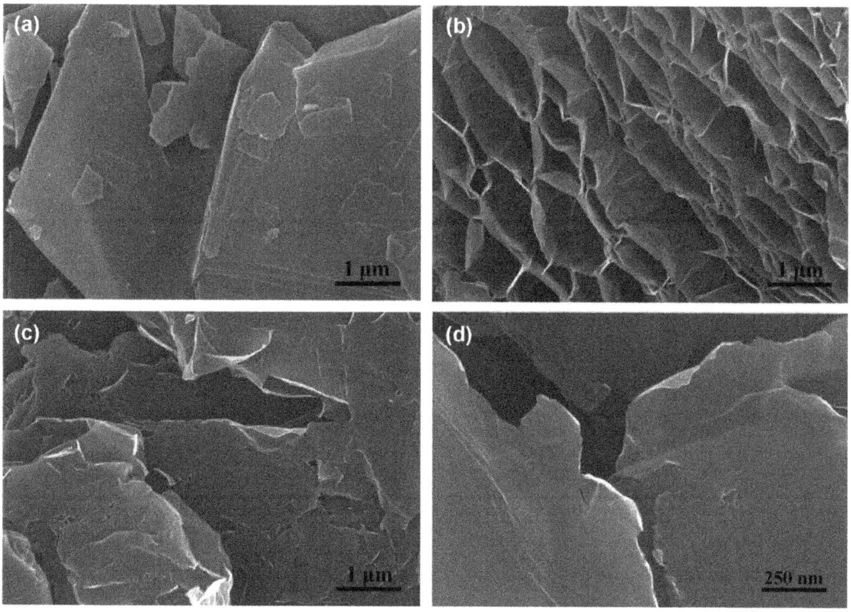

Figure 8.5 SEM images of (a) graphite platelets, (b) graphene oxide, and (c and d) reduced graphene oxide.[60]
Reproduced with permission from ref. 59 with permission from OSA Publishing.

trapped inside the layer. Surface roughness, which can be directly measured by AFM, affects the adsorption and desorption properties of graphene membrane. Membranes with higher roughness tend to have contaminants accumulated in the dents, thus increasing the fouling potential.[63,64] On the other hand, Shukla *et al.* found that a lower roughness and high hydrophilicity of the membrane prevent fouling as well as the adsorption of protein molecules on the surface.

8.3.3.3 Fourier-transform Infrared Spectroscopy (FTIR) and Raman Spectroscopy

FTIR spectroscopy and Raman spectroscopy differ in many ways; firstly, Raman spectroscopy detects the polarizability of a molecule, yet FTIR spectroscopy depends on the change of dipole moment. By measuring the sample's scattered radiation, Raman spectroscopy obtains relative frequencies, but FTIR spectroscopy obtains absolute frequencies by measuring the absorbed radiation of samples.

Both methods can be used to analyze the composition of materials and give them a unique fingerprint: the vibration modes of molecules can be characterized. Investigating GO membranes with Raman spectroscopy indicates the oxidized functional groups,[65] which have an increased density in graphene oxide. The work of Gupta *et al.*[66] demonstrated that the D-band scattering intensity (~1350 cm^{-1}) decreases with increasing layers of graphene. Furthermore, studies have testified that those peaks relate to the microstructure of graphene-based materials.[67] It is also common to use D-band scattering to determine the crystallinity and hence the quality of carbon-based membranes since the D band is known as the disorder band or the defect band.

8.3.3.4 X-Ray Diffraction (XRD)

By diffracting an X-ray beam in different directions, the crystallinity, crystallite size as well as interlayer spacing of graphene-based membranes can be revealed. For the water desalination, water purification, molecule sieving and gas separation properties, interlayer spacing plays the most important role. For instance, graphene oxide in both the dry and humidified state shows an interlayer spacing regulated permeation rate of gas molecules and solvated ions.[39]

XRD can also investigate the change of interlayer spacing of graphene oxide under all circumstances; in the presence of oxygen functional groups,[39] the interlayer spacing can be expanded to 0.9 nm or more in an ionic solution. This can be helpful in understanding the manipulation of GO spacing that is crucial and still challenging for mass transport graphene-based membranes.

8.3.3.5 Other Measurements of Mass Transport Graphene-based Membranes

8.3.3.5.1 Contact Angle. Contact angle, in other words, the wetting properties, influences the water permeability in a different way.[63] Designed to enhance the dispersity of graphene membrane in water solution, graphene oxide membrane is expected to have a lower contact angle than graphene, which means the surface of graphene oxide is amphiphilic. G. Singh and co-workers[68] demonstrated that a high wetting surface can improve the water permeability of graphene oxide membranes to up to 126%. Fast water permeance leads to efficient mass transport and filtration.

8.3.3.5.2 Zeta Potential. Known as the electrokinetic potential, zeta potential is caused by the adsorption of charged particles on a membrane surface. Since graphene family membranes have a high specific surface area, their adsorption properties are crucial for their application in a liquid environment.[22,69] Furthermore, a negatively high zeta potential prevents graphene oxide nanosheets from aggregating in their solvated form.[70]

With the development of interest in reduced graphene oxide membrane, zeta potential is used as circumstantial evidence of the reduction and modification of graphene oxide, because once the number of oxygen groups on the graphene surface increases, zeta potential will drop due to the decreased polarization of the graphene oxide surface.[19]

8.3.3.5.3 Permeability/Selectivity. In comparison with the upper bound polymer membranes, the selectivity–permeability graph is one of the major characterization methods of gas separation properties. Gas permeance is expressed in GPUs, $1 GPU = 1 \cdot 10^{-6} cm^3 cm^{-2} sec\, cmHg$ STP.

Most permeabilities of gases are in inverse proportion to their molecular diameter, but not for methane. The permeability of methane through wetted graphene oxide membrane is higher than carbon dioxide, while its molecular size is 0.38 nm, higher than carbon dioxide (0.33 nm).[71] Furthermore, the permeance of water gas drops with temperature at around 50 °C, but for most other gases, permeance rises because of an increase of oxidation caused defects.[72] Investigation of permeance and selectivity gives access to mechanisms of mass transport of membranes.

8.3.4 Properties of Graphene Oxide

It has been suggested that the top layer of GO membrane is less dense than its core layer as the layers are assembled with less water flow in vacuum filtration. But, this does not affect its mechanical properties compared with different thickness membranes.[56] On varying the thickness of the membrane, the ratio of top layer to the total thickness varies.[56] The top layer may lead to a dominant effect on the mechanical properties for a low thickness membrane while it has an insignificant effect on large thickness membranes. On comparing the mechanical strength of different thicknesses of GO membranes, no noticeable effect is observed.[56]

The mechanical strength of GO film is very high.[46] From the experimental results of Dikin *et al.*, the average tensile modulus out of 31 samples of GO film is 32 GPa.[56] The ultimate tensile elongation without exhibiting slip–stick behavior is 0.6%.[56] Another experimental result from Nair *et al.* also revealed that GO film, which has a thickness smaller than one micron, could withstand a differential pressure of more than 100 mbar.[46]

Note that the mechanical properties of GO membranes vary with the water content. The modulus increases from 17 GPa to 25 GPa with increasing temperature from 40 °C to 120 °C.[56] This is because water content decreases with increased temperature. The lower the water content, the higher the modulus.

Also, the mechanical and electrical properties of GO on EPD were investigated by Kang *et al.*[73] From their work, the hardness and Young's modulus of thin film GO were determined by nanoindentation with a Berkovich diamond probe. Samples of different thicknesses were examined. The values of the mechanical properties were measured from the loading and unloading curve. Their results showed that the stiffness, Young's modulus, and sheet resistance are significantly affected by the film thickness. The larger

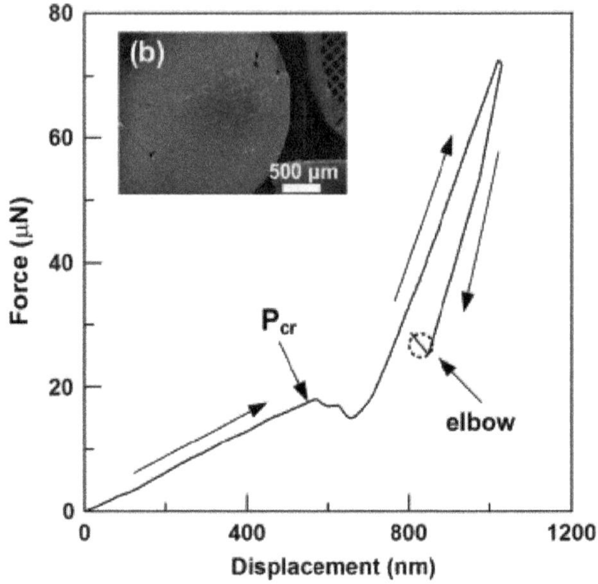

Figure 8.6 The indentation cycle loading and unloading curve showing non-complete recovery and elbow.[73]
Reproduced from ref. 72, with permission of Springer Nature, Copyright © 2013, Springer Nature.

the film thickness,[73] the better the mechanical performance. And, the electrical conductivity is adversely proportional to the film thickness.

During the loading deformation, the curves have linear slopes. And, the stresses show a small drop beyond a critical load. Then, the curves keep rising linearly until the maximum tensile strength is reached. During unloading, elbow curves are observed near the end of unloading for low loads. For high loads, the curves may not completely go back with large deformation during loading.[73] Figure 8.6 below from Kang *et al.*[73] shows the results of mechanical strength of a GO film obtained by conducting an indentation test.

Dikin *et al.*[62] also measured the density of graphene oxide sheets by Archimedes' principle in water.[62] The result obtained was $\approx 1.8\ \mathrm{g\,cm^{-3}}$. Note that the density of GO varies with the different preparation methods and preparation conditions.

8.3.5 Graphene Oxide Membranes as Ion and Molecule Sieves

The permeation of GO membranes has been studied by several groups. One of the research groups used a U-shaped tube in which the feed and the permeate are separated by a GO membrane. The GO membrane was fabricated by stacking multiple layers of GO to form GO laminates. Figure 8.7

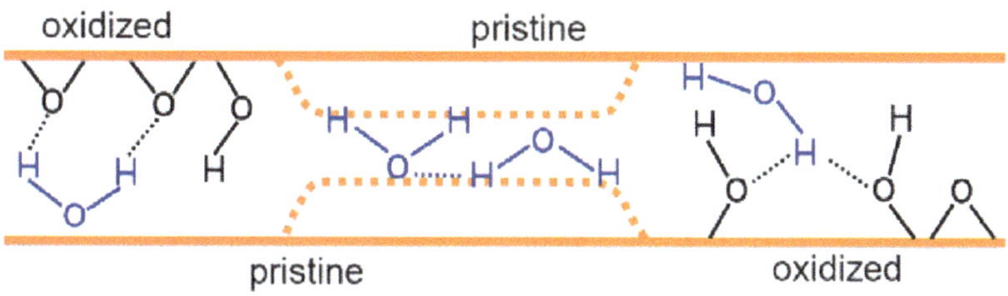

Figure 8.7 Schematic model of the nanocapillary formed between the layers of GO.[42]
Reprinted from *Applied Materials Today*, **1**, R. K. Joshi, S. Alwarappan, M. Yoshimura, V. Sahajwalla and Y. Nishina, Graphene oxide: the new membrane material, 1–12, Copyright 2015, with permission from Elsevier.

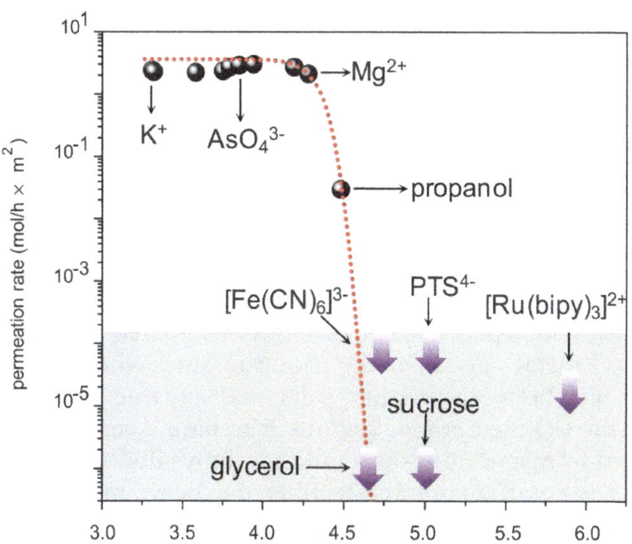

Figure 8.8 Permeation rate of different hydrated ions showing a sharp cutoff at 4.5 Å.[39]
Reproduced from ref. 39, reprinted with permission from AAAS.

below shows the mechanisms of molecules passing through the GO membrane and the experimental setup used in their filtration test.[39]

GO membranes have been identified as permeable membranes for selective gas transport and molecular sieving or absorption. Joshi *et al.*[39] conducted a study of the permeation of GO membranes by filling different liquids into the feed compartment, and the results are shown in Figure 8.8. A rapid drop in the permeation rate is observed at a radius of about 4.5 Å. In other words, GO membrane can act as a molecular sieve allowing ions with hydrated radii smaller than 4.5 Å to pass through.[39] Moreover, they reported that the ion permeation occurs only in the presence of water and the permeation rate increases with water potential.[39] The permeation of small ions is facilitated by a large capillary force owing to the high absorption efficiency of graphite oxide with respect to the salts.[39] This results in ultrafast ion permeation in GO membranes. The schematic model of the nanocapillary is shown in Figure 8.7. Nair *et al.*[40] suggested that a monolayer and two layers of water formed between the graphene sheets, which expand the interlayer spacing up to 10 Å with little friction inside the GO nanocapillaries. This theory was used to explain the ultrafast permeation of water through GO membrane.[40]

Furthermore, it is noteworthy that ion charge does not affect the permeation rate. Joshi *et al.*[39] demonstrated that no significant differences in permeation rate are observed between charged ions such as AsO_43^-, Na^+, and Cl^-.[39] However, experimental results from another group of researchers found that the permeation is dependent on the ionic charge or molecule charge in the feed solutions.[72] Huang *et al.*[72] explained that the negatively charged carboxyl groups attached to the edge of the GO flakes could effectively trap the positively charged molecules. Thus, the pores become partially blocked and permeation rate decreases.[72] Moreover, Hu and Mi found that the permeability is strongly related to the particle charge Debye lengths, or electrostatic repulsion between ions and the membrane charge.[55,74] The controversy in permeability and permselectivity with respect to the ion charge needs further study. Sun *et al.*[55] reported that the permeation rate changes dramatically in different metallic salt solutions. The heavy metal ions permeate much slower compared to the sodium salts because heavy metal salts block the nanocapillaries when passing through the pore.[55] Furthermore, copper salts and organic contaminants are blocked by the GO membrane. Sodium salts have a very high permeation rate, followed by magnesium salts and cadmium salts. The result indicates that GO can act as a membrane to filter out heavy metals from nuclear wastes.[55]

The rejection is related to the number of GO layers. Hu and Mi demonstrated that 5–25 GO layers can reject 10–20% NaCl,[74] as shown in Figure 8.9. GO membranes of 15 and 20 layers show a higher level of NaCl rejection compared to those with 5 and 7 layers of GO. However, no significant trend was observed when Na_2SO_4 and methylene blue (MB) were used. Rhodamine WT (R-WT) has a rejection of ~90% with increasing number of layers.

8.3.6 Graphene Oxide Membranes for Gas Separation

Gas separation has been considered as a crucial step in various important industries like natural gas purification, carbon dioxide sequestration, hydrogen production and oxygen combustion.[70] Membrane separation systems offer great opportunities compared with conventional cryogenic and sorption-based methods, especially in inorganic gas separation applications.[70] The low-energy consumption, no-exotic chemical,

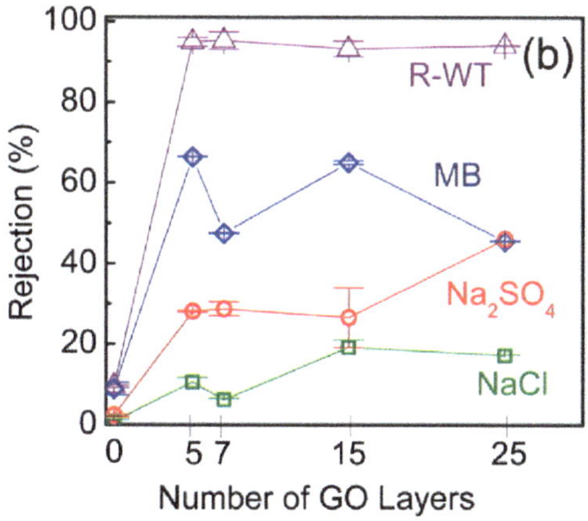

Figure 8.9 The plot of rejection for NaCl with number of GO layers.[74] Reprinted with permission from *Journal of Physics and Chemistry of Solids*, **74**, S.-H. Kang, T.-H. Fang and Z.-H. Hong, Electrical and mechanical properties of graphene oxide on flexible substrate, 1783–1793, Copyright 2013, with permission from Elsevier.

configurationally flexible membrane separation system is a rising star in future industrial mass transportation. However, the trade-off of permeability and selectivity of polymer-based membranes,[13] and the low mechanical capability of zeolite-based membranes[75] have restrained the gas separation membranes from industrial systems in chemically active and stress loaded environments.

Graphene oxide membranes, as a derivative of graphene, have a relatively high Young's modulus and chemical resistance.[29,73] Their permeation and selectivity have been demonstrated to be as high as modified zeolite membranes and silicon carbide membranes.[71] Moreover, as a 2-dimensional material, both the interlayer structure and in-plane pore size of graphene oxide membranes have the potential of being designed through chemical and physical approaches, which gives a promising future to GO membranes for gas separation application.

Although the mechanism of the separation properties of graphene oxide membranes remains unclear, the considerable selectivity of 300 for H_2/CO_2 has attracted the attention of researchers.[76,77] Since the discovery of Nair *et al.*[40] that helium is not permeable for 1 μm thick graphene oxide membrane while graphene oxide membrane shows high selectivity for hydrogen and other gases, several theories have attempted an explanation. M. Yu *et al.*[71] demonstrated that thickness is crucial for gas separation; the permeance of hydrogen and helium is approximately 10 times higher when the graphene oxide membrane is much thinner. As shown in Figure 8.10, they also discovered the relation of permeance of a gas through nanometer-thin graphene oxide membrane and kinetic diameter. Helium and hydrogen have the highest permeance (about 300 times higher than CO_2), and the reason

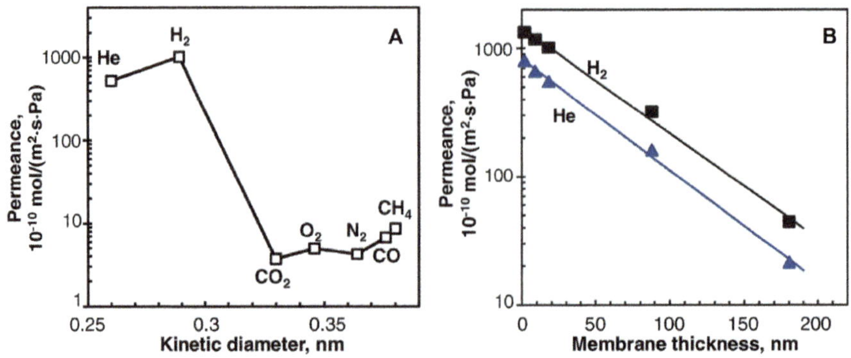

Figure 8.10 The permeation of gases through GO membrane at 20 °C. (a) Variation of permeation against the gas kinetic diameter is plotted. (b) Relationship between permeation of gases (He and H) and membrane thickness.[71]
Reprinted with permission from ref. 70, Copyright 2014 the American Chemical Society.

implied but not testified is that their kinetic diameter is lower than the probed pore size.

As we discussed in the section on characterization technology, gas separation requires not only high permeability but also high selectivity. M. Yu and co-workers[71] tested the selectivity of 1.8, 9.0, and 18 nm thick graphene oxide membranes and found that 9.0 nm thick graphene oxide membrane had the highest selectivity around 900 at 20 °C. This can be demonstrated by simulation results from Rohit Karnik *et al.*[70] that the pore size and resistance of the support material also has an impact on selectivity. When the intrinsic defects are relatively small, as Figure 8.11 demonstrates, high selectivity can be achieved for any support pore size. However, when the film only consists of a few layers, the pore size of the support material should be smaller than the intrinsic defect spacing (L_1 and L_2). So, the intrinsic defects will be the minority of the supporting pores, leaving a majority of support pores with high selectivity.

8.3.7 Graphene Oxide for Desalination

To date, reverse osmosis (RO) has been one of the most extensively industrialized desalination technologies worldwide.[78–80] RO, based on traditional polymeric membranes, is the most energy-efficient desalination technique, which achieves 1.8 kWh m^{-3} compared to the value of 5 kWh m^{-3} in the 1990s.[81] However, the working output of current technologies accounts for only a small fraction of the world's potable water supply, which means that the desalination cost is still high and it consumes a large amount of energy.[82,83] Also, as reported by the International Water Management Institute, water scarcity, the lack of sufficient available water resource to meet

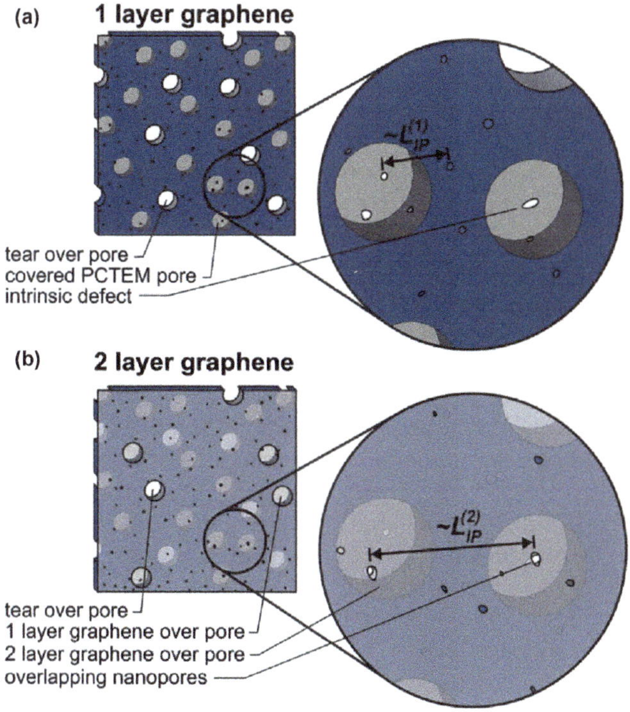

Figure 8.11 (a) Single layer and (b) double layer graphene with intrinsic defects on a porous support material.[70]
Reproduced from ref. 69, with permission from Springer Nature, Copyright 2016, Springer Nature.

water needs, will be one of the most critical global crises by 2025. Thus, it is vital to upgrade the current water desalination/purification system, seek alternative candidates, or/and develop new systems.

Graphene oxide membrane has been identified as a promising candidate for nanofiltration membranes because of its potential for ultrafast water flux and selective permeation, excellent mechanical strength, and chemical inertness.[37–40] The preparation of graphene oxide membranes is cost-effective regarding raw materials and methodologies compared to pristine graphene and carbon nanotubes.[84] Additionally, oxygenated functional groups (such as carboxyl, hydroxyl and carbonyl groups, *etc.*) offer a bonus to the graphene oxide of high hydrophilicity and dispersity.[47,84] Mishra *et al.*[84] reported that the desalination of seawater and simultaneous removal of sodium and inorganic arsenic substances were achieved by a functionalized graphene sheet based water filter.[84]

Besides the inexpensive, thermally and chemically robust character of graphene oxide, the charge selectivity and scalable properties also enable applications in high-performance ion exchange and electro-dialysis technologies. S. Garaj[85] and co-workers testified the charge-selectivity behavior of graphene oxide membrane, when applied with 40 mV of electro-field, the diffusion can be balanced with a drift process. Thus, $FeCl_3$ and $MgCl_2$ ions can be prevented from penetrating through the membrane (Figure 8.12). The influences of pH are further discussed in the article, which provides comprehension of solvated graphene oxide.

Biofouling has been one of the major problems of water desalination for membranes applied in both water and energy industries. Researchers combined graphene oxide membranes with anti-organism materials to produce high permeability and selectivity films to prevent biofouling and improve water flux at the same time. H. Wang and co-workers[86] established a polymer network composited graphene oxide membrane for water desalination, with a highly cross-linking structure and chlorine resistance. As Figure 8.13 shows, the poly(NIPAM-MBA) frame is synthesized along with graphene oxide flakes on a porous substrate; the chlorine tolerance is dramatically increased after 1000 hours of treatment, with water flux stable at as high as 20 $Lm^2h.^2$

Further, it was shown by another research group that the graphene oxide on polytetrafluoroethylene (PTFE) membrane significantly enhanced the water flux (97 $kg\,m^{-2}$) with complete salt rejection.[87] The enhancement might be attributed to the selective sorption, nanocapillary effect, and reduced temperature polarization and the existence of the oxygen functional groups. Overall, the functionalized graphene sheets could possibility be used not only in industrial wastewater disposal but also for desalination.

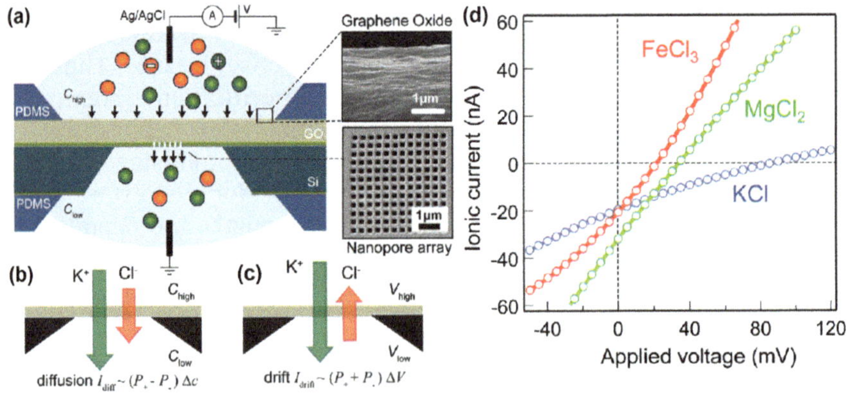

Figure 8.12 (a) Schematics of the experimental setup. (b and c) Depiction of the ionic flow across the membrane driven by the concentration gradient and voltage difference. (d) Ionic current–voltage characteristics of the membrane for different salts.[85]
Reprinted with permission from *Desalination*, **282**, A. K. Mishra and S. Ramaprabhu, Functionalized graphene sheets for arsenic removal and desalination of seawater, 39–45, Copyright 2011, with permission from Elsevier.

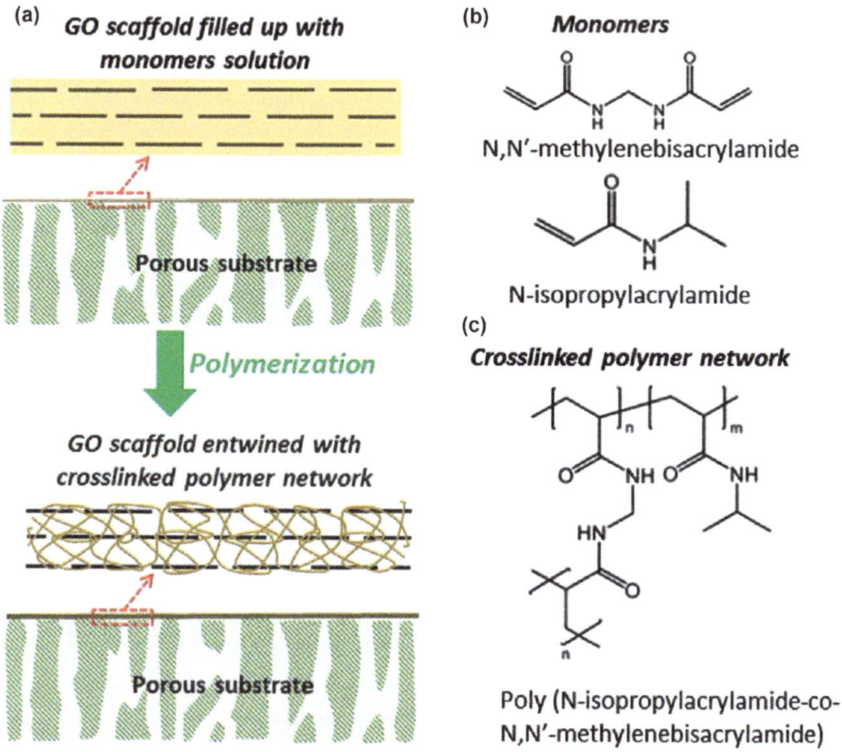

Figure 8.13 (a) Schematic diagram of the fabrication of a GO–polymer network composite membrane. (b) Molecular structure of monomers. (c) Highly crosslinked polymer network formed in the composite membrane.[86] Reprinted with permission from ref. 85, Copyright 2017 the American Chemical Society.

8.4 Summary

In this chapter, we discussed various applications of graphene-based membranes. We clearly pointed out several advantages of graphene-based membranes over other inorganic and polymer membranes. We conclude that graphene-based membranes have huge potential for water purification, desalination and wastewater treatment.

References

1. J. Wang, J. Zhu, Y. Zhang, J. Liu and B. Van der Bruggen, *Nanoscale*, 2017, **9**, 2942–2957.
2. Y. Zhang, X. Feng, S. Yuan, J. Zhou and B. Wang, *Inorg. Chem. Front.*, 2016, **3**, 896–909.
3. N. Jusoh, Y. Fong Yeong, T. Leng Chew, K. Keong Lau and A. Mohd Shariff, *Sep. Purif. Rev.*, 2016, **454**, 321–344.

4. E. Adatoz, A. K. Avci and S. Keskin, *Sep. Purif. Technol.*, 2015, **152**, 207–237.
5. S. Krause, V. Bon, I. Senkovska, U. Stoeck, D. Wallacher, D. M. Többens, S. Zander, R. S. Pillai, G. Maurin, F.-X. Coudert and S. Kaskel, *Nature*, 2016, **532**, 348–352.
6. V. M. A. Melgar, J. Kim and M. R. Othman, *J. Ind. Eng. Chem.*, 2015, **28**, 1–15.
7. R. Castro-Muñoz, V. Fíla and C. T. Dung, *Chem. Eng. Commun.*, 2017, **204**, 295–309.
8. B. Li, H. Wang and B. Chen, *Chem. – Asian J.*, 2014, **9**, 1474–1498.
9. S. T. Muntha, A. Kausar and M. Siddiq, *Polym. -Plast. Technol. Eng.*, 2016, **55**, 1282–1298.
10. G. Liu, W. Jin and N. Xu, *Angew. Chem., Int. Ed.*, 2016, **55**, 13384–13397.
11. Z. Lai, G. Bonilla, I. Diaz, J. G. Nery, K. Sujaoti, M. A. Amat, E. Kokkoli, O. Terasaki, R. W. Thompson, M. Tsapatsis and D. G. Vlachos, *Science*, 2003, **300**, 456–460.
12. A. Corma, F. Rey, S. Valencia, J. L. Jordá and J. Rius, *Nat. Mater.*, 2003, **2**, 493–497.
13. L. M. Robeson, *J. Membr. Sci.*, 2008, **320**, 390–400.
14. D. F. Sanders, Z. P. Smith, R. Guo, L. M. Robeson, J. E. McGrath, D. R. Paul and B. D. Freeman, *Polymer*, 2013, **54**, 4729–4761.
15. J. Lee, H. R. Chae, Y. J. Won, K. Lee, C. H. Lee, H. H. Lee, I. C. Kim and J. Min Lee, *J. Membr. Sci.*, 2013, **448**, 223–230.
16. J. Shen, G. Liu, K. Huang, Z. Chu, W. Jin and N. Xu, *ACS Nano*, 2016, **10**, 3398–3409.
17. Z. Zhang, H. T. H. Nguyen, S. A. Miller, A. M. Ploskonka, J. B. Decoste and S. M. Cohen, *J. Am. Chem. Soc.*, 2016, **138**, 920–925.
18. H. Bruschke, *Pure Appl. Chem.*, 1995, **67**, 993–1002.
19. L. C. Lin and J. C. Grossman, *Nat. Commun.*, 2015, **6**, 8335.
20. K. C. Khulbe, T. Matsuura, C. Y. Feng and A. F. Ismail, *RSC Adv.*, 2016, **6**, 42943–42961.
21. H. Li, K. Haas-Santo, U. Schygulla and R. Dittmeyer, *Chem. Eng. Sci.*, 2015, **127**, 401–417.
22. C. A. Scholes and U. K. Ghosh, *Membranes*, 2017, 7, 1–13.
23. A. Livingston, L. Peeva, S. Han, D. Nair, S. S. Luthra, L. S. White, L. M. Freitas and D. Santos, *Ann. N. Y. Acad. Sci.*, 2003, **984**, 123–141.
24. D. D. L. Chung, *J. Mater. Sci.*, 2002, **37**, 1475–1489.
25. H. C. Schniepp, J. L. Li, M. J. McAllister, H. Sai, M. Herrera-Alonson, D. H. Adamson, R. K. Prud'homme, R. Car, D. A. Seville and I. A. Aksay, *J. Phys. Chem. B*, 2006, **110**, 8535–8539.
26. S. Gilje, R. B. Kaner, G. G. Wallace, D. A. N. Li, M. B. Mu, M. B. Muller, S. Gilje, R. B. Kaner and G. G. Wallace, *Nat. Nanotechnol.*, 2008, **3**, 101–105.
27. M. Kim, N. S. Safron, E. Han, M. S. Arnold and P. Gopalan, *Nano Lett.*, 2010, **10**, 1125–1131.
28. I. Ovid'ko, *Rev. Adv. Mater. Sci.*, 2013, **34**, 1–11.

29. L. Yan, Y. B. Zheng, F. Zhao, S. Li, X. Gao, B. Xu, P. S. Weiss and Y. Zhao, *Chem. Soc. Rev.*, 2012, **41**, 97–114.
30. M. Ulbricht, *Polymer*, 2006, **47**, 2217–2262.
31. M. D. Fischbein, M. Drndi, M. D. Fischbein and M. Drndi, *Appl. Phys. Lett.*, 2011, **93**, 113107.
32. S. Garaj, W. Hubbard, A. Reina, J. Kong, D. Branton and J. A. Golovchenko, *Nature*, 2010, **467**, 190–193.
33. D. C. Bell, M. C. Lemme, L. A. Stern, J. R. Williams and C. M. Marcus, *Nanotechnology*, 2009, **20**, 455301.
34. D. Cohen-Tanugi and J. C. Grossman, *Nano Lett.*, 2012, **12**, 3602–3608.
35. K. Sint, B. Y. Wang and P. Kral, *J. Am. Chem. Soc.*, 2009, **131**, 9600.
36. Y. Zhu, S. Murali, W. Cai, X. Li, J. W. Suk, J. R. Potts and R. S. Ruoff, *Adv. Mater.*, 2010, **22**, 3906–3924.
37. B. Mi, *Science*, 2014, **343**, 740–742.
38. Y. Han, Z. Xu and C. Gao, *Adv. Funct. Mater.*, 2013, **23**, 3693–3700.
39. R. K. Joshi, P. Carbone, F. C. Wang, V. G. Kravets, Y. Su, I. V. Grigorieva, H. A. Wu, A. K. Geim and R. R. Nair, *Science*, 2014, **343**, 752–754.
40. R. R. Nair, H. A. Wu, P. N. Jayaram, I. V. Grigorieva and A. K. Geim, *Science*, 2012, **335**, 442–444.
41. N. Wei, C. Lv and Z. Xu, *Langmuir*, 2014, **30**, 3572–3578.
42. R. K. Joshi, S. Alwarappan, M. Yoshimura, V. Sahajwalla and Y. Nishina, *Appl. Mater. Today*, 2015, **1**, 1–12.
43. S. Stankovich, D. A. Dikin, R. D. Piner, K. A. Kohlhaas, A. Kleinhammes, Y. Jia, Y. Wu, S. B. T. Nguyen and R. S. Ruoff, *Carbon*, 2007, **45**, 1558–1565.
44. D. W. Boukhvalov, M. I. Katsnelson and Y. W. Son, *Nano Lett.*, 2013, **13**, 3930–3935.
45. A. V. Talyzin, V. L. Solozhenko, O. O. Kurakevych, T. Szabó, I. Dékány, A. Kurnosov and V. Dmitriev, *Angew. Chemie - Int. Ed.*, 2008, **47**, 8268–8271.
46. R. R. Nair, H. A. Wu, P. N. Jayaram, I. V. Grigorieva and A. K. Geim, *Science*, 2012, **335**, 442–444.
47. W. Gao, in *Graphene Oxide: Reduction Recipes, Spectroscopy, and Applications*, 2015, pp. 61–95.
48. L. Stobinski, B. Lesiak, A. Malolepszy, M. Mazurkiewicz, B. Mierzwa, J. Zemek, P. Jiricek and I. Bieloshapka, *J. Electron Spectrosc. Relat. Phenom.*, 2014, **195**, 145–154.
49. Y. You, X. H. Jin, X. Y. Wen, V. Sahajwalla, V. Chen, H. Bustamante and R. K. Joshi, *Carbon*, 2018, **129**, 415–419.
50. H. L. Poh, F. Šaněk, A. Ambrosi, G. Zhao, Z. Sofer and M. Pumera, *Nanoscale*, 2012, **4**, 3515.
51. W. S. Hummers and R. E. Offeman, *J. Am. Chem. Soc.*, 1958, **80**, 1339.
52. M. W. S. Gilje and S. Han, *Nano Lett.*, 2007, **7**, 3394–3398.
53. J. Chen, B. Yao, C. Li and G. Shi, *Carbon*, 2013, **64**, 225–229.
54. C. Botas, P. Álvarez, P. Blanco, M. Granda, C. Blanco, R. Santamaría, L. J. Romasanta, R. Verdejo, M. A. López-Manchado and R. Menéndez, *Carbon*, 2013, **65**, 156–164.

55. P. Sun, M. Zhu, K. Wang, M. Zhong, J. Wei, D. Wu, Z. Xu and H. Zhu, *ACS Nano*, 2013, 7, 428–437.
56. J. Song, X. Wang and C.-T. Chang, *J. Nanomater.*, 2007, **448**, 457–460.
57. S. Stankovich, D. A. Dikin, G. H. B. Dommett, K. M. Kohlhaas, E. J. Zimney, E. A. Stach, R. D. Piner, S. T. Nguyen and R. S. Ruoff, *Nature*, 2006, **442**, 282–286.
58. M. Hu and B. Mi, *J. Membr. Sci.*, 2014, **469**, 80–87.
59. J. Boguslawski, J. Sotor, G. Sobon, R. Kozinski, K. Librant, M. Aksienionek, L. Lipinska and K. M. Abramski, *Photonics Res.*, 2015, **3**, 119–124.
60. H. Feng, X. Wang and D. Wu, *Ind. Eng. Chem. Res.*, 2013, **52**, 10160–10171.
61. A. C. Society and S. Xi-scientific, *Atomic and Electronic Structure of Solids*, 2011.
62. I. Jung, M. Pelton, R. Piner, D. A. Dikin, S. Stankovich, S. Watcharotone, M. Hausner and R. S. Ruoff, *Nano Lett.*, 2007, 7, 3569–3575.
63. E. Yuliwati and A. F. Ismail, *Desalination*, 2011, **273**, 226–234.
64. T. H. Bae, I. C. Kim and T. M. Tak, *J. Membr. Sci.*, 2006, **275**, 1–5.
65. D. Yang, A. Velamakanni, G. Bozoklu, S. Park, M. Stoller, R. D. Piner, S. Stankovich, I. Jung, D. A. Field, C. A. Ventrice and R. S. Ruoff, *Carbon*, 2009, **47**, 145–152.
66. A. Gupta, G. Chen, P. Joshi, S. Tadigadapa and P. C. Eklund, *Nano Lett.*, 2006, **6**, 2667–2673.
67. K. N. Kudin, B. Ozbas, H. C. Schniepp, R. K. Prud'homme, I. A. Aksay, R. Car, R. K. Prud'homme, I. A. Aksay, R. Car, R. K. Prud'homme, I. A. Aksay and R. Car, *Nano Lett.*, 2007, **8**, 36–41.
68. J. A. Prince, S. Bhuvana, V. Anbharasi, N. Ayyanar, K. V. K. Boodhoo and G. Singh, *J. Membr. Sci.*, 2016, **500**, 76–85.
69. S. M. Fatemi and M. Foroutan, *Int. J. Environ. Sci. Technol.*, 2016, **13**, 457–470.
70. M. S. H. Boutilier, C. Sun, S. C. O'Hern, H. Au, N. G. Hadjiconstantinou and R. Karnik, *ACS Nano*, 2014, **8**, 841–849.
71. H. Li, Z. Song, X. Zhang, Y. Huang, S. Li and Y. Mao, *Science*, 2013, **342**, 95–98.
72. H. Huang, Z. Song, N. Wei, L. Shi, Y. Mao, Y. Ying, L. Sun, Z. Xu and X. Peng, *Nat. Commun.*, 2013, **4**, 2979.
73. S.-H. Kang, T.-H. Fang and Z.-H. Hong, *J. Phys. Chem. Solids*, 2013, **74**, 1783–1793.
74. M. Hu and B. Mi, *Environ. Sci. Technol.*, 2013, **47**, 3715–3723.
75. N. Kosinov, J. Gascon, F. Kapteijn and E. J. M. Hensen, *J. Membr. Sci.*, 2016, **499**, 65–79.
76. S. P. Koenig, L. Wang, J. Pellegrino and J. S. Bunch, *Nat. Nanotechnol.*, 2012, **7**, 728–732.
77. H. W. Kim, H. W. Yoon, S. Yoon, B. M. Yoo, B. K. Ahn, Y. H. Cho, H. J. Shin, H. Yang, U. Paik and S. Kwon, *Science*, 2013, **342**, 91–96.

78. C. Fritzmann, J. Löwenberg, T. Wintgens and T. Melin, *Desalination*, 2007, **216**, 1–76.
79. D. Li and H. Wang, *J. Mater. Chem.*, 2010, **20**, 4551.
80. H. T. El-Dessouky, H. M. Ettouney and Y. Al-Roumi, *Chem. Eng. J.*, 1999, **73**, 173–190.
81. M. Elimelech and W. A. Phillip, *Science*, 2011, **333**, 712–718.
82. I. C. Karagiannis and P. G. Soldatos, *Desalination*, 2008, **223**, 448–456.
83. C. Charcosset, *Desalination*, 2009, **245**, 214–231.
84. A. K. Mishra and S. Ramaprabhu, *Desalination*, 2011, **282**, 39–45.
85. S. Hong, C. Constans, M. V. Surmani Martins, Y. C. Seow, J. A. Guevara Carri and S. Garaj, *Nano Lett.*, 2017, **17**, 728–732.
86. S. Kim, X. Lin, R. Ou, H. Liu, X. Zhang, G. P. Simon, C. D. Easton and H. Wang, *J. Mater. Chem. A*, 2017, **5**, 1533–1540.
87. M. Bhadra, S. Roy and S. Mitra, *Desalination*, 2016, **378**, 37–43.

CHAPTER 9

Graphene-based Membranes for Water Desalination Applications

P. S. GOH AND A. F. ISMAIL*

Advanced Membrane Technology Research Centre (AMTEC), Faculty of Chemical and Energy Engineering, Universiti Teknologi Malaysia, 81310, Johor Bahru, Malaysia
*Email: fauzi.ismail@gmail.com

9.1 Introduction

The water shortage issue can be mainly attributed to rapid industrialization, global population growth and increased urbanization.[1] Water desalination has played an increasingly crucial role in resolving the water scarcity problem to meet ever growing demands. Particularly, it has been addressed as a sustainable approach to supply constant clean water to drought areas such as the Middle East and areas where the freshwater supply is limited such as Singapore. Desalination has been feasibly applied for a wide range of sources including seawater, brackish water and wastewater. While desalination of seawater or brackish water has been generally accepted as a feasible alternative to freshwater production, the energy demands of the existing desalination technologies in terms of high grade thermal energy and/or prime quality electricity have constantly posed challenges in their practical applications.[2] Compared to thermal based desalination that is known to be more energy prohibitive, membrane-based desalination technologies with lower energy consumption have served as a more attractive option in

Nanoscience & Nanotechnology Series No. 47
Graphene-based Membranes for Mass Transport Applications
Edited by Pengzhan Sun and Hongwei Zhu

Published by the Royal Society of Chemistry, www.rsc.org

many coastal and metropolitan cities to support the water demands. Currently, membrane desalination comprises about 65% of the total global desalination capacity of ~100 $M m^3$ per day with a total market of ~\$30 billion in 2015.[3]

Since its first commercialization in the 1970s, reverse osmosis (RO) has been widely applied, and it is one of the fastest growing water treatment technologies.[4] Although the improvement in energy recovery and membrane materials made in the past few decades has allowed improvements in desalination economics, efforts have been constantly made to further enhance the technology and system efficiency to reduce the environmental impact and energy usage.[5] Minimizing energy consumption has the greatest prospect to reduce overall desalination costs. Forward osmosis (FO) is an alternative desalination process that could serve as an option for reducing the energy requirements. In brief, FO is an osmotically driven membrane process that uses the advantage of the osmotic pressure gradient to drive the permeation of water across the membrane, hence it holds promise to address several shortcomings of the hydraulic pressure-driven membrane RO process such as relatively lower energy consumption as well as lower fouling propensity and higher fouling reversibility.[6] However, the development of an ideal FO membrane with the desired high water flux and rejection toward the feed and draw solutes is still a challenge to be addressed. To achieve higher efficiency, the major parameters of membrane processes, *i.e.* selectivity, permeability and concentration polarization/fouling, should be taken into consideration. The development of sustainable, durable and energy-efficient membranes at commercially attractive cost for both seawater and brackish water desalination is an important task that requires innovative ideas and practical implementations.[7] One of the most straightforward approaches to simultaneously resolve the issues related to these parameters is through the design of a novel membrane. Currently, RO and FO desalination operations strongly rely on the performance of polyamide (PA) thin film composite (TFC) membranes. Despite the improvement made in the past few decades to enhance the polymeric cross-linkage of the PA layer and the overall membrane surface functionalities and morphologies, it was pointed out that the most promising TFC membranes today are only up to 2 folds more permeable than those developed 20 years ago.[8] Moreover, the state-of-the-art TFC membranes are still vulnerable to chlorine attack and highly susceptible to a wide range of fouling. Hence, this class of RO membranes still has much room for improvement. Material scientists and engineers are exploring the potential of nanomaterials that are expected to not only produce novel desalination membranes, but also heighten the performance of the typical desalination technologies.

Nanotechnology, which literally refers to the manipulation of materials and processes in the molecular scale of 1–100 nm, is an emerging field that allows the hierarchical construction of materials to render remarkable performance improvement and add new functionality to the existing products. In the last two decades, a wide range of nanomaterials has been developed and used for aquatic pollutant treatment.[9] Particularly, nanostructured materials have offered unprecedented energy-efficient desalination through improvement in the membrane permeability. For example, owing to the molecular nanotube wall smoothness, aligned carbon nanotubes have demonstrated water fluxes that are three to four orders of magnitude higher than those expected from conventional hydrodynamic theories.[10] Currently, graphene-based nanomaterials are at the forefront of the research to revolutionize next generation RO membranes. Graphene-based membranes possess tremendous potential to surpass the performance of the existing technologies for desalination and ionic or organic molecular separation in aqueous environments due to their well-defined channels and also the presence of oxygen-containing functional groups in some of the graphene derivatives.[11] It was predicted that nanoporous graphene (NPG) would demonstrate 2–3 orders of magnitude higher water permeability compared to that of conventional RO membranes.[12,13] NPG with precisely controlled and high-density subnanometer pores has been proposed as a medium to allow ultrafast water transport while retaining high salt rejection due to the atomic thinness with minimum flow resistance. The overview of the current trends of graphene-based membranes in terms of the fabrication approach, performance and enhancements in desalination and other membrane separation processes has been comprehensively reviewed.[14,15]

This chapter discusses the performance and the development of graphene-based membranes for desalination application. In the first section, the properties of graphene-family nanomaterials and their fabrication methods are covered. The next section discusses the development of graphene-based membranes in both computational and experimental studies. Their applications in several desalination processes such as RO, FO, pervaporation and membrane distillation are thoroughly discussed. The computational work related to graphene-based membranes for desalination processes is also covered. Finally, the concluding remarks and a brief future outlook of the development of graphene-based membrane technologies for desalination are highlighted.

9.2 Properties and Synthesis of Graphene-family Nanomaterials

As illustrated in Figure 9.1, graphene is a two-dimensional material that is made up of covalently bonded carbon atoms, which form a hexagonal honeycomb lattice.[13] Due to the presence of sp^2 hybridized carbon atoms, graphene has been generally depicted as having a flat, planar configuration. In fact, both simulation and experimental studies have suggested that when graphene sheets are freely suspended in air or a vacuum, they are corrugated to some extent due to thermal fluctuation.[16] The corrugation has been identified as the key to controlling the properties of graphene for both fundamental and technological purposes. Graphene is currently known as

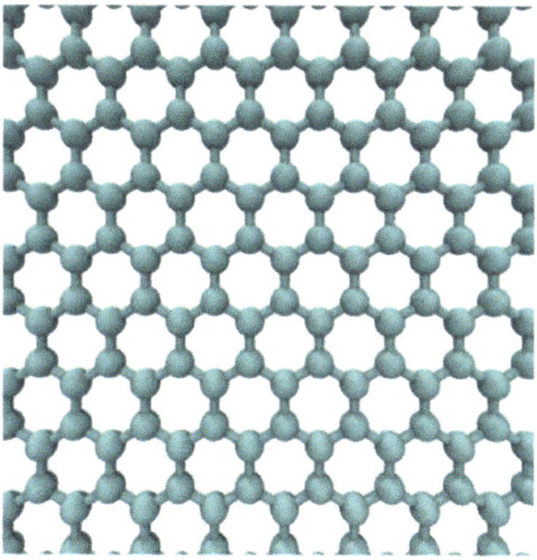

Figure 9.1 Crystal lattice of graphene.
Reprinted from *Desalination*, **366**, D. Cohen-Tanugi and J. C. Grossman, Nanoporous graphene as a reverse osmosis membrane: Recent insights from theory and simulation, 59–70, Copyright 2015, with permission from Elsevier.[13]

one of the thinnest and strongest materials that have been explored. Although much of the original interest was related to graphene's unique electronic properties, the mass separation application of graphene has also been explored in recent years. Graphene holds promise as an attractive RO membrane for water desalination as the atomic thinness of graphene presents a principally simpler desalination process. Through the creation of subnanometer pores on the graphene sheet, the single-layer NPG can be used as a membrane for water treatment.[14] Additionally, owing to the strength and robustness of the graphene structure, NPG can serve as a thin semi-permeable membrane where a fabric support substrate is not needed. These advantages are promising for high membrane permeability to increase the water productivity during the desalination process. The reason for the high water permeability has been attributed to the low-friction water movement across the non-oxidized regions near the graphene surface.[17]

Processes such as chemical etching and block copolymer templating have been explored to form high-density pores of precise diameter size distribution in the graphene monolayers.

O'Hern *et al.* pointed out that sealing of the graphene membrane defects is key to the practical application of NPG for molecular-level filtration.[18] They reported a multiscale leakage–sealing process to selectively seal the defects based on the nonpolar nature and impermeability of pristine graphene. In order to seal the nanometer-scale intrinsic defects created during chemical vapor deposition and the transfer of graphene copper foil to a polycarbonate track etch membrane, a two step procedure, which involved selectively filling with hafnia using atomic layer deposition (ALD) and interfacial polymerization reaction of aqueous hexamethylenediamine (HMDA) and adipoyl chloride (APC) in hexane, was carried out.

Various methods have been established to produce various types of graphene-based nanomaterials. Graphene was first mechanically exfoliated from small mesas of fresh graphite through a simple scotch tape technique from monocrystalline graphitic films by Novoselov *et al.*[19] However, the yield was too low for the large-scale production of graphene. Batch or continuous chemical vapor deposition (CVD) is a widely used approach to produce graphene sheets in higher yields. This approach is favoured for industry application as it allows large-area film production and the films can be feasibly transferred onto a wide range of substrates.[20] In brief, CVD can be carried out *via* several routes, which include thermal, plasma enhanced and thermal decomposition on SiC and other substrates. Many attempts have been made to synthesize mono or multilayered graphene on several types of metal substrates.[21,22] Upon the completion of CVD, etching of the metal substrate is performed to detach the graphene layers in order to transfer them to a new substrate without undergoing complicated mechanical exfoliation or chemical reduction. Kafiah *et al.* reported the transfer of monolayer graphene onto a polypropylene (PP) and polyvinylidenedifluoride (PVDF) microfiltration substrate for water desalination.[23] As illustrated in Figure 9.2, removal of graphene was performed from one side by the graphene/Cu produced *via* CVD over an etchant. The graphene monolayer was then attached to a polymeric substrate membrane by sandwiching them with the graphene/Cu between two glass slides and roll pressing with a glass rod. Finally, the transfer process was completed by using the same etchant to etch away the Cu.

The chemical exfoliation of graphite results in the formation of monolayer GO flakes attached by various forms of oxygen-containing functional groups. GO exhibits similar properties as graphene but the sheets appear to be asymmetrical due to the presence of oxygen-containing functional groups on the edges and basal planes. The presence of epoxy, hydroxyl, carbonyl and carboxylic acid groups allows the stable dispersion of monolayer GO flakes in water. However, thermal, chemical or irradiation reduction is needed to restore the electrical conductivity of monolayer GO. Chemical reduction of GO sheets can be carried out with several reducing agents including hydrazine and sodium borohydrate to produce reduced graphene oxide (rGO).[24] During the conversion process, the reduced sheets aggregated and precipitated, hence becoming less hydrophilic due to the removal of oxygenated functional groups. Structurally, rGO is similar to GO but it is not completely homogenous like graphene due to the presence of remaining functional groups. Thermal reduction has also been reported to reduce GO to rGO. In this approach, heat treatment is applied to eliminate the oxide functional groups from GO surfaces.[25]

The surface functionalization of GO plays an important role in altering the exfoliation behavior of GO and rGO, hence it is the key that leads to various applications. The surface functionalization has been commonly performed

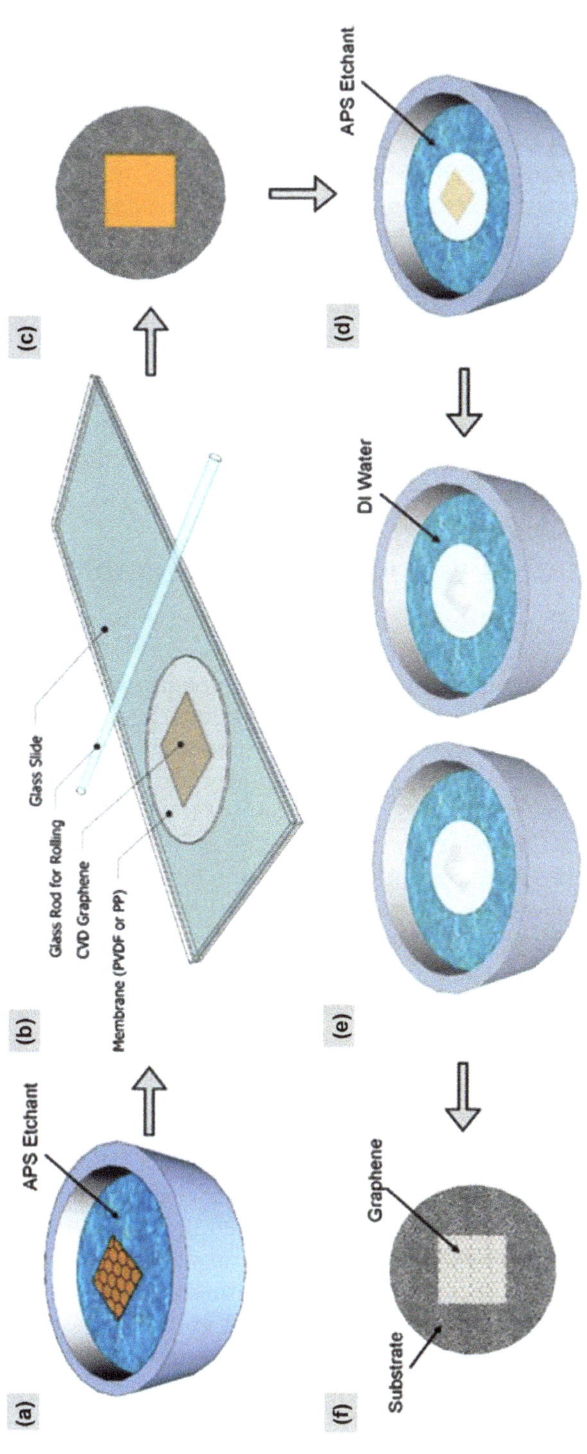

Figure 9.2 Schematic drawing of monolayer graphene transfer onto polymeric substrate. Reprinted from *Desalination*, **388**, F. M. Kafiah, Z. Khan, A. Ibrahim, R. Karnik, M. Atieh and T. Laoui, Monolayer graphene transfer onto polypropylene and polyvinylidenedifluoride microfiltration membranes for water desalination, 29–37, Copyright 2016, with permission from Elsevier.[23]

via two approaches, *i.e.* covalent functionalization and non-covalent functionalization. Covalent functionalization involves the modification of the oxygen functional groups including the carboxylic acid groups at the edges and epoxy/hydroxyl groups on the basal plane on GO surfaces in order to modify the overall surface functionality of GO. On the other hand, non-covalent functionalization of GO is based on the weak π–π interactions, which include van der Waals interactions and electrostatic interaction. The sp^2 network on GO can render π–π interactions with targeted conjugated polymers and aromatic compounds to produce functional composite materials.

Graphene nanoplatelets (GNPs) are normally obtained from the exfoliation of graphite in batches to obtain a large quantity of free-standing or powder-like nanomaterial, which varies in thickness and composition. Another synthesis route of GNPs involves the bottom-up approach, which uses a gas-phase and self-assembly process. The main advantage of this approach is that the hydrophilic characteristic can be introduced in GNPs to facilitate their dispersion.

9.3 Fabrication of Graphene-based Membranes

In general, graphene-based membranes can be prepared in two configurations, *i.e.* free-standing and nanocomposite membranes. One of the possible ways to prepare stacked graphene-based membranes is through the filtration of the graphene sheets through membrane filters.[26] The stacked membranes can then be peeled from the filters after drying in air or vacuum environments. The filtration approach has the advantages of simplicity and high reproducibility while allowing precise control over the thickness of the stacked membranes. Besides that, a gravure printing machine with a high concentration of graphene-based sheets can continuously produce large-area stacked membranes.[27] Despite the exceptional properties shown by the free-standing graphene membranes, technical challenges such as the mechanical strength as well as the thermal and chemical stability are encountered in these free-standing membranes, which further hinder their practical application.

Unlike the free-standing nanoporous single-layer graphene membrane, nanocomposite membranes containing GFN are technically easier to produce hence can be feasibly realized in scaled-up commercialization. A polymeric support substrate or host matrix is commonly used in the fabrication of nanocomposite membranes to provide high mechanical strength as well as high physical and chemical stability. The fabrication of GFN incorporated nanocomposite membranes is generally categorized into two primary approaches. Graphene-based nanomaterials can be introduced onto the membrane surfaces through the post-treatment modification of the membrane's selective layer. The combination of graphene-based nanomaterials and polymeric substrate allows the formation of a new class of nanocomposites that merges different properties in the individual material.

Typically, the graphene-based nanomaterials can be embedded into the selective layers of TFC membranes during the interfacial polymerization (IP) process, which involves the polymerization of two monomers in aqueous and organic phases. Some other methods that have also been attempted include covalent functionalization, the layer-by-layer (LbL) approach and vacuum-assisted filtration.[14] Each of these techniques results in a membrane surface with a different extent of uniformity in thickness and scalability. The incorporation of graphene-based nanomaterials in the substrate layer of the TFC has also been commonly reported. The so-called mixed-matrix membranes were successfully prepared through the physical mixing of GFN with various polymers such as PVDF, polysulfone (PS) and polyethersulfone (PES) during dope formulation.[7] These nanocomposite membranes exhibit a remarkable enhancement in their physicochemical properties, which cannot be attained from the conventional composites. However, it should be noted that the degree of improvement is strongly influenced by the dispersion of functionalized graphene-based nanomaterials in the polymer matrix. Besides that, the previous studies working along this direction have pointed out some common issues to be addressed. Some of them include precise pore control to optimize the flow passage and the establishment of viable routes for the incorporation of well-dispersed graphene-based nanomaterials in an identified host matrix without compromising the overall structural integrity.

Overall, the incentives of incorporating these graphene-based nanomaterials to form the nanocomposite membranes are to facilitate fast water transport within the selective layers or polymer supports, hence improving the membrane flux and productivity of the desalination process. Another important improvement can be associated with the conferral of antimicrobial and chemical resistance properties through the enhanced membrane hydrophilicity, which leads to a greater antifouling property. The membrane morphology and surface roughness can be tailored and fine-tuned to achieve this purpose. Besides that, the introduction of these nanomaterials also provides better mechanical strength in the polymeric substrates to withstand the high pressure applied during the typical RO desalination. It is interesting to note that the desirable membrane properties and the extent of these effects can be engineered through the adoption of different modification approaches.[28]

9.4 Performance of Graphene-based Membranes

9.4.1 Reverse Osmosis

The incorporation of graphene-based nanomaterials into RO semi-permeable membranes has been widely attempted to improve water permeation by increasing the membrane surface hydrophobicity as well to enhance the salt rejection ability by altering the free volumes across the selective layers, hence controlling the diffusion of hydrated ions. The presence of graphene-based nanomaterials is also related to the increase in thermal stability, solvent resistance, lifetime and antibacterial activity of the resultant nanocomposite membranes.[29–31] The incorporation of discrete GO nanosheets within the TFC PA layer through IP was first reported by He *et al.*[32] Compared to the unmodified TFC, the pure water flux across the TFN incorporated with 0.12 wt.% GO was increased by 80% from 0.122 to 0.219 L μm m^{-2} h bar without compromising the salt rejection, owing to the lowered membrane surface energy and enhanced hydrophilicity rendered by the well-dispersed GO. Based on the interesting findings obtained, the authors also proposed that the highly negative zeta potential at high pH of the GO TFN membranes could provide a significant improvement in the antifouling properties and this property is beneficial for the treatment of pH sensitive effluents where the iso-electric point of the membrane can be tailored beyond the limit of unmodified TFC membrane.

Similarly, Chae *et al.* prepared GO TFN membrane by incorporating the chemically exfoliated GO into the PA layer.[33] The fractionated GO was dispersed in m-phenylenediamine (MPD) aqueous solution during the IP process and the RO performance was evaluated in a 2000 ppm NaCl aqueous solution in a cross-flow system. It was found that the GO-incorporated TFN membrane becomes smoother, more hydrophilic and more negatively charged with the increasing GO content. The concerted effects contributed by these enhancements resulted in improved flux as well as chlorine and microbial resistance. The water permeability and anti-biofouling property of the GO-embedded TFN membrane were enhanced by 80% and 98%, respectively. The increased chlorine resistance of the membrane was attributed to hydrogen bonding between GO and PA, which could protect the underlying PA from chlorine attack. Similar findings have also been reported by Choi *et al.*, who applied the LbL assembly of GO multilayers on the PA layer.[34] In order to create a conformal GO assembly in a controllable manner, positively charged aminated-GO (AGO) and negatively charged GO were alternately deposited on the PA layer through electrostatic interaction, as shown in Figure 9.3(a). As shown in Figure 9.3(b) and (c), the membrane surface is significantly smoothened by the GO sheets, which have filled the valley regions of the rough PA surface. Despite the increase of membrane thickness due to the deposition of GO layers, the water flux and salt rejection of the GO deposited TFC membrane remained similar to those of unmodified TFC, which were around 14 L m^{-2} h and 96.4%, respectively. The reduced fouling tendency in the GO modified membrane was ascribed to the enhanced hydrophilicity and the reduced surface roughness, which in turn minimized the preferential attraction between the hydrophobic foulant molecule and the PA surface. Recently, Shao *et al.* also modified the PA layer of RO TFC membrane with GO layers through a spin coating technique.[35] The membrane exhibited good chlorine resistance to prevent membrane degradation as the GO layer rendered protective properties to the PA functional layer by hampering the chlorine radicals from forming O–Cl bonds.

Safarpour *et al.* prepared TFN RO membrane by introducing rGO/TiO_2 hybrid nanomaterials into the PA layer of the TFN through the IP of MPD and TMC.[36] The surface morphology of the PA layer was not significantly

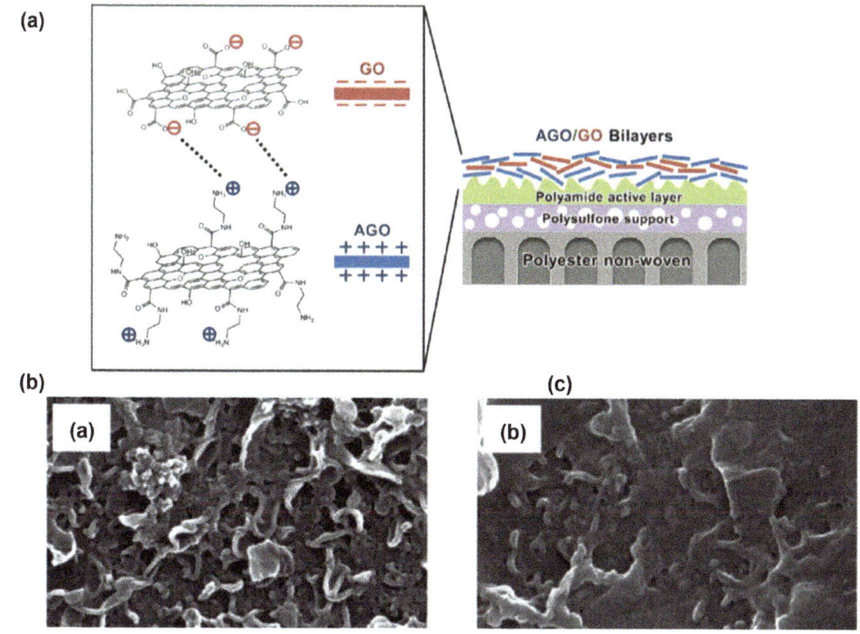

Figure 9.3 (a) Schematic illustration of a multilayered GO LbL coating on a PA TFC. Atomic force microscopy images of (b) GO-coated TFC and (c) unmodified TFC.
Reprinted with permission from W. Choi, J. Choi, J. Bang and J. Lee, *ACS Appl. Mater. Interfaces*, 2013, **5**, 12510–12519, Copyright 2013 American Chemical Society.

altered but the hydrophilicity of the membranes was greatly improved with the increasing concentration of rGO/TiO_2 owing to the existence of different hydrophilic and negatively charged functional groups on the hybrid nanomaterial. The optimum RO performance was achieved with 0.02 wt% rGO/TiO_2 and a water flux of 51.3 $L\,m^{-2}\,h$ and salt rejection of 99.45% were obtained. The trade-off was not observed, showing that the increase of water flux did not result from the membrane defects caused by the addition of the nanomaterials to the PA layer. Additionally, the salt rejection of the TFN RO membrane after chlorination tests barely decreased by 3%, which was 10 times lower than that of neat TFC, indicating the good chlorine resistance of the TFN. Again, the improvement was attributed to the enhanced hydrophilicity and the barrier created by rGO/TiO_2 to protect the PA thin layer from contacting the corrosive chlorine environment.

Instead of using conventional PA RO membranes, which are known to be susceptible to chlorine attack due to their relatively low chemical resistance, Chen *et al.* explored the use of GO modified cellulose triacetate (CTA) as an RO membrane to deal with seawater feed that suffers from severe biofouling.[37] The flat-sheet RO nanocomposite membrane was fabricated with GO embedded within the CTA layer *via* a melting method using a vulcanization machine. The incorporation of GO resulted in the improvement in the mechanical properties of the CTA nanocomposite membrane where the tensile strength increased from 10.2 MPa to 23.1 MPa. This improvement was attributed to the dominant interactions between C=O groups from CTA and –OH and –COOH groups from GO including van der Waals interaction and hydrogen bonding. When tested for RO desalination, the permeate flux of the membrane increased from 1.67 $L m^{-2} h$ to 4.74 $L m^{-2} h$ with the increasing GO content. However, the trade-off effect was observed where the salt rejection was reduced with the increased permeate flux. This phenomenon was ascribed to the formation of channels that act as a passageway for not only water but also salt ions when the GO content was increased from 0.02 wt% to 0.1 wt%.

Prior to the incorporation of GO in the PA layer of RO membranes, Kim *et al.* performed the surface modification of GO using self-polymerization of tannic acid in basic buffer solution on the GO surface.[38] Due to the fast water transport on the GO surface, a water flux of 9.00 $L m^{-2} h$ was observed for the RO membrane embedded with 0.05 wt% functionalized GO, which represented a 26.3% increment compared to the neat PA membrane, while maintaining the salt rejection ability of above 95%. Another significant improvement was related to the chlorine resistance properties in which the PA RO membrane containing tannic acid functionalized GO exhibited 0.94 normalized salt rejection even at 35 000 ppm h of chlorine exposure. The excellent anti-microbial performance of RO membrane incorporated with tannic acid functionalized GO was attributed to several advantageous properties of the nanofiller in terms of its improved barrier properties, oxidative stress capability and radical scavenging features contributed by both GO and tannic acid. GO quantum dots (GOQDs), normally with the diameter range of 3–20 nm, are known for their unique properties in terms of superhydrophilicity, ultra-small lateral sizes and cytotoxicity; hence, they can be used as a potential candidate for anti-(bio) fouling agents in water purification membranes. Fathizadeh *et al.* first reported the incorporation of nitrogen-doped GO quantum dots (N-GOQDs) into the PA layer of RO membranes during the IP process.[39] The N-GOQDs showed good compatibility with the PA layers as the terminal amine groups could favourably form chemical bonds with the PA matrix. Furthermore, upon the reaction of N-GOQDs with MPD and TMC monomers, the crosslinking formed could create larger transport pathways at the interface due to the relatively wide distribution of functional groups on the N-GOQDs. At low loadings of N-GOQDs, these channels could effectively facilitate the water transport. The TFN membrane exhibited a water permeability of 0.72 $L m^{-2} h$ and a salt rejection of 93.1%, which represented a 3-times increase of water permeability compared to that of neat TFC membranes.

9.4.2 Forward Osmosis

FO has attracted growing attention in desalination due to its advantages over the conventional RO desalination in terms of low energy consumption and higher antifouling ability as the operation of FO does not require external hydraulic pressure. However, one of the issues facing FO is the internal concentration polarization (ICP) that tends to reduce the water-permeation flux by more than 80%. Subsequently, huge efforts have been made to reduce the ICP in order to improve the FO efficiency for practical usage. To fully utilize the advantages of GO to increase the surface hydrophilicity and render anti-microbial effects, Shen *et al.* incorporated GO nanosheets into the PA rejection layer to develop TFN membrane for FO application. Compared to the nanomaterials that are buried within the bulk polymeric supports, the incorporation of GO into the PA layer allows the exposure of the nanomaterials to the feed solution. They observed that the GO-incorporated PA layer was endowed with enhanced hydrophilicity and surface smoothness. A thinner PA selective layer was also formed due to the higher reaction rate between TMC and MPD with the addition of GO. As a result, the TFN membranes possessed desired characteristics such as higher water permeability and salt rejection, lower salt permeability and improved antifouling properties as compared to the control TFC membrane. The highest flux recorded was 35.4 $L\,m^{-2}\,h$. The water flux deterioration of the GO-incorporated TFN membranes was less severe and the flux recovery ratio exceeded 90%. However, when the amount of GO exceeded the optimum loading of 600 ppm, the agglomeration of GO occurred and led to pinhole formation on the membrane surface. Hence, several solutions that involved the process of GO synthesis and the preparation of GO-incorporated TFN membranes have been proposed, *i.e.* (i) increase the oxidation degree of the GO to obtain a smaller particle size, (ii) surface modification of GO to promote dispersibility in the aqueous solution and (iii) exfoliation of GO by high frequency ultrasonication to break the agglomeration. Salehi *et al.* reported the synthesis of high-performance chitosan/graphene oxide incorporated FO membranes with tailored composition and tunable permeability *via* layer-by-layer (LbL) assembly.[40] Chitosan was used to crosslink GO nanosheets on the negatively charged sulfonated PES/PES substrate. Electrostatic interactions induced by the oppositely charged substances could improve the interfacial compatibility between active layer and substrate. Compared to the TFC, the LbL membranes showed improved water flux without showing much effect on the salt permeability when tested with 1 M Na_2SO_4 or 1 M sucrose aqueous solutions as the draw solution and deionized water as the feed solution. The presence of GO can enable low-friction flow of water molecules due to the hydrophobic nature of the graphitic nanochannels where the hydrogen bonds between water molecules promote the ultrafast transport through the two-dimensional capillaries.

Hegab *et al.* attempted the incorporation of GO nanosheets onto the PA layer of FO TFN membranes using bioadhesive polydopamine (pDA).[41] The self-assembly and oxidative polymerization of pDA reduced and immobilized GO nanosheets to form a durable, flexible and nanoscale layer on the membrane surface. The coating of GO nanosheets, as illustrated in Figure 9.4(a), involved the crystallization and restacking of the mixture of

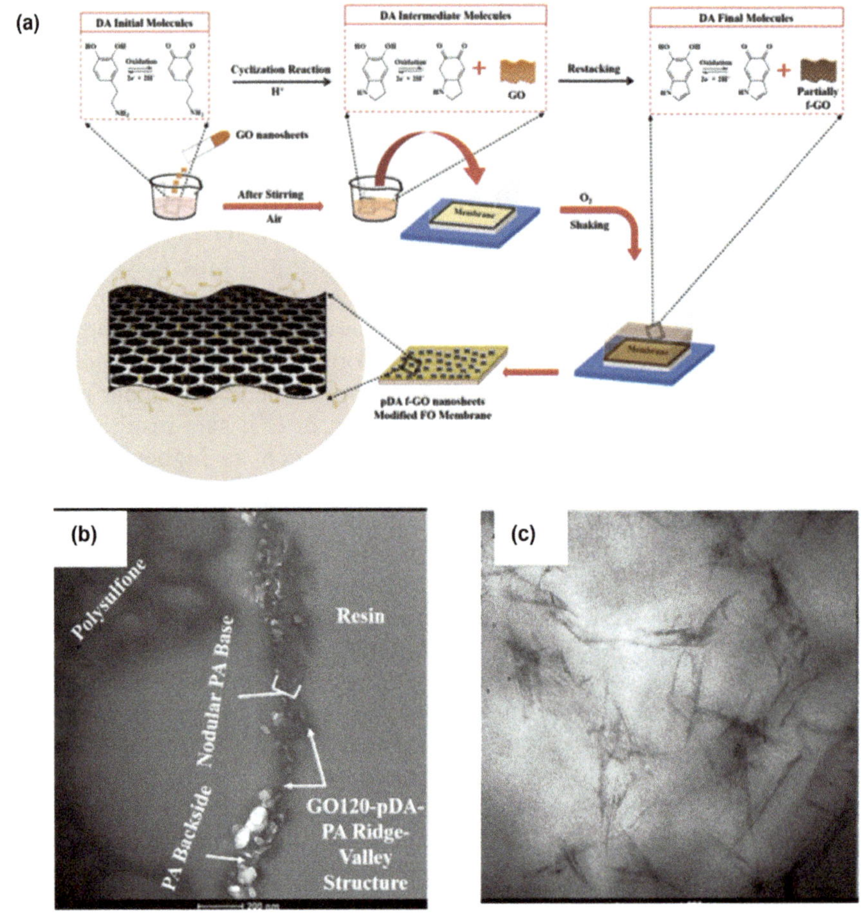

Figure 9.4 (a) Schematic illustration of GO–pDA *in situ* membrane surface modification technique. Transition electron microscope images of (b) neat and (c) GO–pDA TFN membranes.
Reprinted from *Desalination*, **385**, H. M. Hegab, A. ElMekawy, T. G. Barclay, A. Michelmore, L. Zou, C. P. Saint and M. Gininc-Markovic, Effective in-situ chemical surface modification of forward osmosis membranes with polydopamine induced graphene oxide for biofouling mitigation, 126–137, Copyright 2016, with permission from Elsevier.[41]

dopamine and graphene nanosheets, followed by the coating onto the FO in an oxygen atmosphere. A GO concentration of 80 $\mu g\,mL^{-1}$ and dopamine deposition time of 30 min represented the optimal surface modification process for the GO modified FO membranes. The incorporation of both GO nanosheets and pDA onto the membrane surface led to the formation of a tight active layer, as shown in Figure 9.4(b). As a result, the reverse solute flux selectivity performance was improved and the Js parameter was reduced by 80% compared to that of the neat membrane.

A similar pDA coating approach was also reported by Yang *et al.* to reduce GO laminates onto a commercial mixed cellulose ester (MCE) membrane. It was found that the rGO laminates could retain their compacted nanochannels with a dimension of 3.45 Å, consequently increasing the selectivity of hydrated ions. The pDA coating played an important role in improving the hydrophilicity of the rGO laminate surface where water could be facilitated to flow into the nanochannels formed in rGO laminates, hence speeding up the water absorption. Owing to the synergistic effects, the MCE FO membrane coated with pDA reduced GO achieved a remarkable improvement in water flux without compromising the salt rejection. The highest water flux and the lowest reverse solute flux obtained were 36.6 $L\,m^{-2}\,h$ and 0.042 $mol\,L\,m^{-2}$, respectively, when tested in FO mode using deionized water as feed solution and 0.6 M NaCl solution as draw solution. Wang *et al.* reported the preparation of rGO modified graphitic carbon nitride (g-C_3N_4), CN/rGO, incorporated porous PES substrate for the fabrication of PA FO membranes. The special architecture and physicochemical properties of CN/rGO in terms of the porosity and the lamellar morphology created by the curved nanosheets were utilized to improve the FO performance. The findings showed that the CN/rGO incorporation could positively affect the permeability of the substrate by decreasing the hydraulic resistance of the membrane. The nanocomposite membrane embedded with 0.5 wt% CN/rGO in the PES of TFC membrane achieved an osmotic water flux of 41.4 $L\,m^{-2}\,h$ when 2M NaCl was used as draw solution and deionized water was used as feed. This flux represented an improvement that was about 20% higher than that of neat TFC membrane. The enhanced water permeability was attributed to the change in the pore structure and enhanced wettability, which eventually resulted in the corresponding smaller structural parameter.

The FO desalination performance of free-standing CTA membrane incorporated with GO was explored by Wang *et al.*[42] The main advantage of this free-standing CTA is the associated low water resistance and ICP due to the absence of a support layer in the membranes. Furthermore, the incorporation of GO increased the hydrophilicity of the CTA membrane. The water flux significantly increased to 18.43 $L\,m^{-2}\,h$ when DI water and 0.5 M NaCl were used as the feed and draw solution, respectively. A low reverse salt flux of 0.22 $g\,L^{-1}$ was also observed. However, it was observed that the water flux behavior was influenced by both the hydrophilicity and hindrance effect of GO. At GO loadings above 0.6 wt%, a drastic decrease in the flux was observed. This phenomenon was due to barrier effects of the two-dimensional graphene sheets in the membrane matrix when they were added in excess. As a result, the water channels were blocked and deflected.

9.4.3 Membrane Distillation

From the membrane distillation application point of view, graphene, which consists of 2D sheets of sp^2-bonded carbon atoms in hexagonal honeycomb lattices, is beneficial to serve as effective sorption sites for water vapors freed from hydrogen bonding but it repels water molecules. It is also agreed that the presence of polar hydroxyl and carboxyl functional groups on GO can change the membrane–water vapor interactions, hence improving the overall vapor permeation rates. Bhadra *et al.* explored the potential of GO membranes in membrane distillation based on the attractive properties of GO in terms of the selective sorption of water vapors, nanocapillary effect for selective sieving of pure water from brine and the reduction in temperature polarization.[43] Additionally, the presence of functional groups in hydroxylated or carboxylated GO is also expected to enhance the flux performance. Through the immobilization of GO on the polytetrafluoroethylene (PTFE) membrane surface, they found that the overall permeate flux could reach as high as 97 $kg\,m^{-2}$ h at 80 °C without compromising the salt rejection. As shown in Figure 9.5, the functional groups tend to distribute and form nanoscale wrinkles and structural defects to offer favourable passages for water transport. On the other hand, the hydrophobic graphene stacks offer a near-frictionless surface for fast water transport and minimize the temperature polarization without resulting in conductive heat losses. These cumulative effects contributed to the flux in membrane distillation.

Graphene-loaded electrospun polyvinylidene fluoride-co-hexafluoropropylene (PH) nanofiber membrane was fabricated for desalination *via* air gap membrane distillation (AGMD). The incorporation of graphene provides value-added properties to the nanofibers by increasing the roughness and hydrophobicity to form a robust and highly efficient membrane distillation membrane. With the incorporation of graphene, a superhydrophobic membrane with porosity of more than 88% and a high liquid entry pressure (LEP) of 4186 kPa was obtained. These favourable properties led to a high and stable AGMD flux of 22.9 $L\,m^{-2}$ h or LMH, which is an improvement of 5 times compared to that of commercial PVDF flat-sheet membrane. Excellent salt rejection of 100% was also observed for 60 h of operation using 3.5 wt% NaCl solution as feed.

9.4.4 Pervaporation

Pervaporation is a chemical potential driven process that is different from RO in terms of the vapor phase present in the permeate stream of pervaporation. GFNs have strong potential to deal with the insufficient flux obtained during the separation and hence they are able to overcome the main drawback of using pervaporation on a commercial scale. GNP incorporated polyether block amide (PEBA) membrane was reported for seawater desalination through pervaporation in a temperature range of 35–65 °C.[44] Good compatibility between the nanomaterial and polymeric phases was obtained owing to the high aspect ratio and high surface area of GNPs. The best salt rejection and flux performance were obtained as 99.94% and 2.58 $kg\,m^{-2}$ h, respectively, at 35 °C when 3 wt% GNPs was incorporated into the hydrophobic PEBA membrane. The long-term stability test indicated

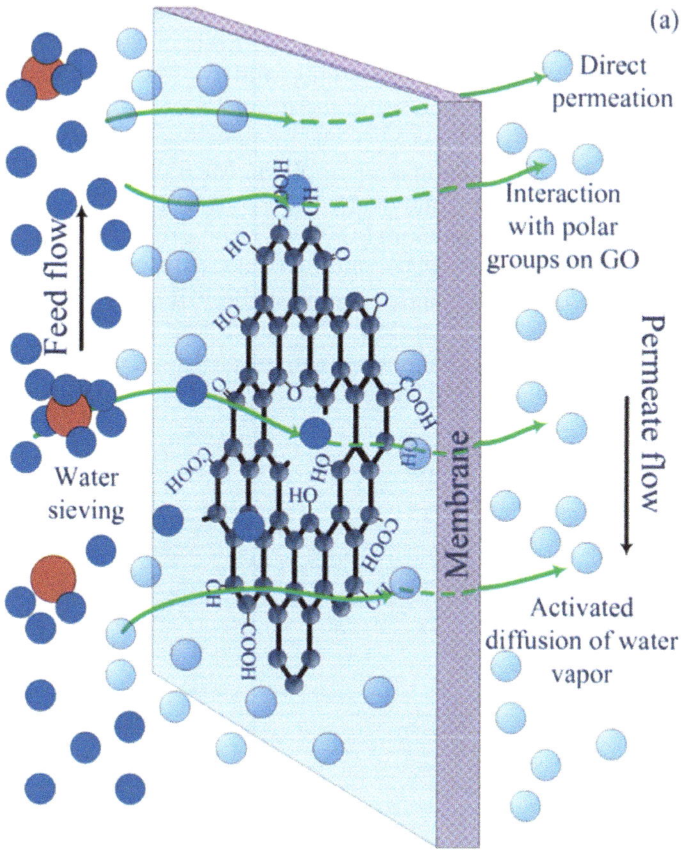

Figure 9.5 Schematic of proposed mechanism of GO membrane in membrane distillation.
Reprinted from *Desalination*, **378**, M. Bhadra, S. Roy and S. Mitra, Desalination across a graphene oxide membrane via direct contact membrane distillation, 37–43, Copyright 2016, with permission from Elsevier.[43]

that the separation performance and lifespan of GNP incorporated PEBA membranes were significantly prolonged due to the anti-microbial properties and high polarity of the GNPs. The addition of hydrophilic GNPs switched the hydrophobicity of PEBA to hydrophilic, hence reducing the accumulation of organic substances on the surface of the membrane.

GO membrane supported on inorganic supports such as Al_2O_3 tube also demonstrated promising results in pervaporation desalination. However, it was found that the GO membranes prepared on porous Al_2O_3 supports tend to have deteriorating performance over time due to crinkling and peeling off of the GO layer from the support. In order to deal with this issue, Xu *et al.* used pDA as a covalent linker to improve the interfacial adhesion and establish better interaction of GO layers with the support surface.[45] The abundant oxygen-containing groups in the GO framework can favourably react with the amine groups of pDA, hence binding the GO nanosheets onto the Al_2O_3 surface for the synthesis of dense and stable GO membranes. With the improved stability and the desired characteristics rendered by GO, the membrane exhibited a high water flux of 48.4 kg m^{-2} h and high ion rejections of over 99.7% in pervaporation desalination of 3.5 wt% seawater at 90 °C. Feng *et al.* applied a vacuum filtration technique to prepare graphene oxide framework (GOF) membrane.[46] As depicted in Figure 9.6(a), the three-dimensional GOF was first prepared by crosslinking 1,4-phenylene diisocyanate (PDI) onto GO nanosheets, followed by the vacuum filtration of the GOF suspension on the PDI-modified α-Al_2O_3 tube to form micron-thick GOF membrane. A dense and defect-free GOF layer with a wavy–wrinkle morphology was formed on the PDI-modified α-Al_2O_3 tube, as shown in Figure 9.6(b). When tested for pervaporation desalination,

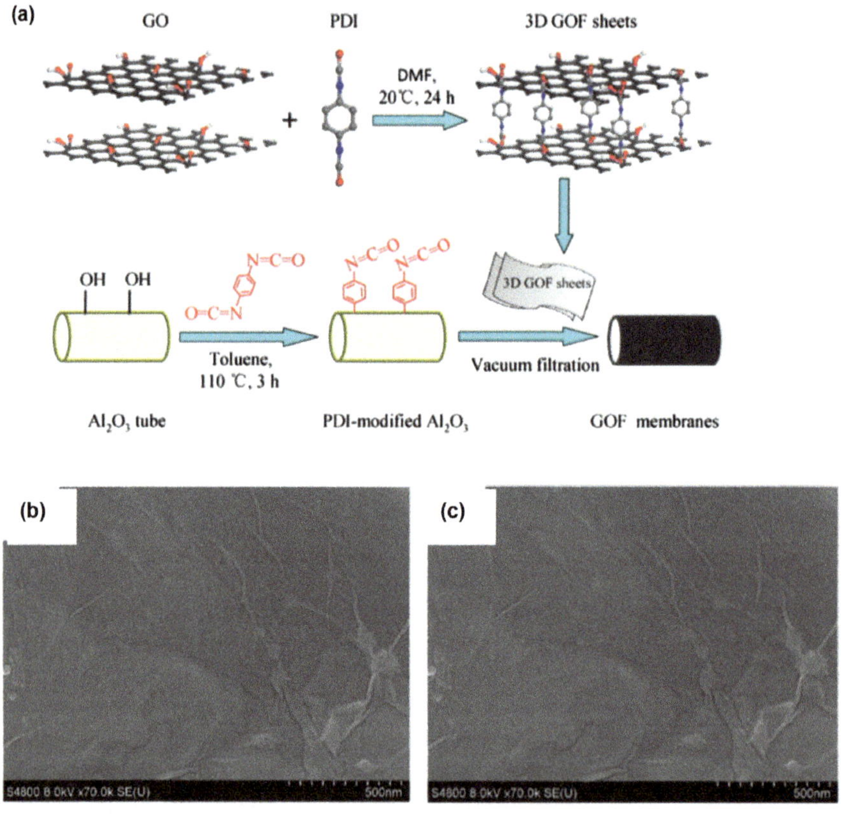

Figure 9.6 (a) Synthesis of GOF membranes on the PDI-modified α-Al_2O_3 tubes through the vacuum filtration method. (b) Top view and (c) cross-section scanning electron microscope images of the GOF membranes. Reprinted from *Desalination*, **394**, B. Feng, K. Xu and A. Huang, Covalent synthesis of three-dimensional graphene oxide framework (GOF) membrane for seawater desalination, 123–130, Copyright 2016, with permission from Elsevier.[46]

it was found that the resistance of the mass transferring through the GOF membrane was significantly reduced due to the three-dimensional structure of the GOF, which in turn resulted in a high water flux of 11.4 $kg\,m^{-2}\,h$ and ion rejection of over 99.9% using 3.5 wt.% seawater at 90 °C. The GOF membrane also demonstrated long-term stability over a long period of operation at 75 °C for 120 h. The authors also proposed that the GOF membranes can be potentially used to treat highly concentrated salt solutions up to 10 wt%.

9.4.5 Computational Studies

Computational studies, spanning from the atomic scale to the macroscopic modeling, can be ideally used to study the mechanisms of water transport to design a radically new concept of desalination membrane. Particularly, Cohen-Tanugi and Grossman performed a decent number of computational works based on molecular dynamics (MD) to promote the applications of graphene membranes for desalination. MD simulations are known to be an ideal tool to study the desalination system using graphene-based membranes as they allow the thorough kinetic and thermodynamic studies of desalination while considering the physics of water, ions, and graphene layers with high accuracy.[47]

The water permeability of NPG under the operating pressure of a real RO system was reported.[48] By using hydrogenated NPG with an open pore area of 23 $Å^2$, the effect of pressure on the water transport in the vicinity of the nanopore was investigated in their MD simulation work. They demonstrated that the NPG pore density of $1.7 \times 10^{13}\,cm^{-2}$ could enable an ultrafast water permeability of approximately $10^3\,L\,m^{-2}\,h\,bar$, which was an order of magnitude greater than that of commercial TFC membranes. However, the adverse effect of concentration polarization, which is associated with the flux decline due to the accumulation of salt near the membrane, was not addressed in this study. The mechanical strength of NPG membranes under the high hydraulic pressure applied in RO desalination was also investigated using MD simulations and continuum fracture mechanics.[49] The authors reported that the size and spacing of the nanopores, as well as the pore size of the support layer played significant roles in affecting the structural resilience of NPGs. Hence, by tailoring the chemistry of the polymeric substrate, the mechanical stability of the NPG membrane can be maintained. The substrate with openings of less than 1 μm was able to maintain the structural integrity at pressure as high as 57 MPa, which is approximately 10 times more than the pressure applied in typical seawater RO desalination processes.

The performance of a GO nanoporous layer as a FO membrane has also been reported based on MD simulations with NaCl used as the feed solution and NH_4HCO_3 used as the draw solution.[50] The authors reported a GO membrane water flux that was about three orders of magnitude greater than that of the commercial cellulose triacetate FO membrane. The GO membrane with a pore diameter of 9.4 Å and porosity of 10% could achieve a water flux of 91.5 L cm^{-2} per day. The enhancement was mainly attributed to the presence of oxygen-containing groups on the surface of GO that rendered the membranes with remarkable hydrophilic features. Zero internal concentration polarization and no reverse solutes across the membrane were observed with the GO membranes, which indicated the potential opportunities of applying functionalized porous GO for seawater desalination as well as power generation through the osmotically driven membrane processes.

Functionalization of nanoporous graphene is an interesting topic for exploration.[51,52] The pore formation in graphene tends to reduce its thermodynamic stability. As such, functional groups are favourably introduced at the pore edge unsaturated carbon atoms of graphene to minimize the formation energy in order to improve the stability of the pore edge. Moreover, pore functionalization is also a promising approach to modify the pore size and pore chemistry, which might in turn improve the permeation performance. Nitrogen atom with comparable atomic size has been agreed as a promising element for the chemical doping of carbon materials. Chen and Yang investigated the separation performance of pyridinic-like nitrogen-doped nanoporous graphene as a RO desalination membrane through molecular simulation study.[53] The nitrogen functionalized graphene membranes allowed moderate free energy barriers for water passage and hence promised greater water fluxes. The water flux reported in this MD study was six times higher than that of pristine graphene and was three times higher than that for CNT membrane. Besides that, the pore area also imparted an important effect on the flux in which the broadening cross-section area in the nitrogen-doped graphene membrane can facilitate faster water passage. On the other hand, it was also found that the salt rejection of the graphene membranes is based on the pore size exclusion of hydrated ions.

9.5 Challenges and Concluding Remarks

Graphene-based nanomaterials represent one of the greatest breakthroughs in technological and industrial development, and they have provoked significant innovations to revolutionize century-old water treatment processes. Particularly, graphene-based membranes are attractive candidates to offer promise for the future of desalination technology. The key to successful industrial application of graphene-based membranes for commercial scale desalination processes strongly relies on the synthesis of economically scalable graphene and graphene-enabled membranes. Significant improvement has been made in the graphene production methods to fulfil the industry expectations in term of its quality, quantity and reproducibility. It is known that the high salt rejection reported in graphene-based membrane was associated with the use of suspended, monocrystalline, single-layer graphene with a small effective membrane surface area in order to avoid the exposure of the domain boundaries of graphene and reduce the defect formation. This has in turn highlighted the remarkable technical challenges that need to be resolved in order to up scale the fabrication of these single-layer graphene membranes for practical desalination applications.

Despite the huge efforts made in synthesizing large-scale graphene films, the fabrication of intact monolayer graphene in large effective surface areas for practical application remains highly challenging experimentally. Currently, the major challenge in graphene membrane synthesis is associated with the stability and reproducibility of the membrane. Despite the ideal structures demonstrated in typical computational studies, it is known that the functional groups at the nanopore are susceptible to deformation under real operating conditions, which was found to be detrimental to the permeability and salt rejection performance of the graphene membrane.[48] Like most of the nanomaterials, one of the key challenges is to address the aggregation issues during the synthesis and processing of bulk-quantity graphene sheets.[54] Due to their structural properties, graphene tends to form agglomerates and restack to form graphite through van der Waals interactions. As the unique properties of GFN are majorly associated with individual sheets, the prevention of aggregation is crucial to attain the desired characteristics.

The most commonly applied technique for the fabrication of free-standing graphene sheets, chemical vapor deposition, is flawed with some inevitable Stone-Wales and other intrinsic defects, hence monolayer regions rarely exceed 1 mm^2.[47] Additionally, the presence of these defects tends to increase the surface chemical activity and consequently disrupt the sp^2-bonding nature of graphene, which was found to be detrimental to the fundamental properties of graphene.[55] Furthermore, although the advances in computational studies allow the understanding of specific physical and chemical properties of ions in a hydrated environment and the complex hydrogen-bonding behavior of water at the nanoscale, graphene-based nanomaterials remain a highly challenging material to be experimentally characterized at the atomic scale. Despite their common application, the above-mentioned chemical reduction approaches to obtain GO have some disadvantages in terms of toxic waste production that is harmful to the environment. For instance, hydrazine vapors are known to be highly poisonous. Additionally, thermal reduction involves heat release and multi-step removal processes with several functional group molecules, which resulted in negative impacts to the environment and economic feasibility when scale-up for production is considered. Hence, more environment-friendly and highly effective reducing agents should be explored to substitute the conventional methods to reduce GO.

Some simulation studies also indicated that the choice of substrate is also a critical parameter to be taken into consideration while designing a practical graphene-based membrane.[49] Despite the above-mentioned efforts, it is also very important to standardize the definition of the members of GNFs. By reassuring the nature and quality of the diverse range of graphene-based nanomaterials, the adoption and commercialization of graphene-enabled technology can be accelerated. Currently, the technology readiness level of graphene-based membranes for desalination is still in the stage of proof-of-concept and lab validation.[56] Nevertheless, the exponential growth in the number of scholarly publications on graphene-enabled desalination and in the number of patents filed indicates the vast potential to expedite the commercialization of graphene-based membranes. It is anticipated that with the availability of graphene-based nanomaterials in the desired quality, amount and cost to meet industrial applications, desalination technology based on graphene-based membranes will serve as a promising candidate to propel development in the water community.

Acknowledgements

The author would like to acknowledge the financial support provided by the Ministry of Higher Education under HiCOE Grant (Vot number 4J182).

References

1. V. G. Gude, *Water Res.*, 2016, **89**, 87–106.
2. J. Schallenberg-Rodríguez, J. M. Veza and A. Blanco-Marigorta, *Renewable Sustainable Energy Rev.*, 2014, **40**, 741–748.
3. J. Imbrogno, J. J. Keating, J. Kilduff and G. Belfort, *Desalination*, 2017, **401**, 68–87.
4. R. L. Stover, *Desalin. Water Treat.*, 2016, **3994**, 1–6.
5. M. Elimelech and W. A. Phillip, *Science*, 2011, **333**, 712–717.
6. D. L. Shaffer, J. R. Werber, H. Jaramillo, S. Lin and M. Elimelech, *Desalination*, 2015, **356**, 271–284.
7. P. S. Goh and A. F. Ismail, *J. Chem. Technol. Biotechnol.*, 2015, **90**, 971–980.
8. K. P. Lee, T. C. Arnot and D. Mattia, *J. Membr. Sci.*, 2011, **370**, 1–22.
9. C. Santhosh, V. Velmurugan, G. Jacob, S. K. Jeong, A. N. Grace and A. Bhatnagar, *Chem. Eng. J.*, 2016, **306**, 1116–1137.
10. B. J. Hinds, N. Chopra, T. Rantell, R. Andrews, V. Gavalas and L. G. Bachas, *Science*, 2004, **303**, 62–65.
11. Z. Liu, W. Wang, X. Ju, R. Xie and L. Chu, *Chin. J. Chem. Eng.*, 2017, **25**, 598–1605.
12. D. Cohen-Tanugi and J. C. Grossman, *Nano Lett.*, 2012, **12**, 3602–3608.
13. D. Cohen-Tanugi and J. C. Grossman, *Desalination*, 2015, **366**, 59–70.
14. K. Goh, H. E. Karahan, L. Wei, T. H. Bae, A. G. Fane, R. Wang and Y. Chen, *Carbon N. Y.*, 2016, **109**, 694–710.
15. P. S. Goh and A. F. Ismail, *Desalination*, 2015, **356**, 115–128.
16. W. Bao, F. Miao, Z. Chen, H. Zhang, W. Jang, C. Dames and C. N. Lau, *Nat. Nanotechnol.*, 2009, **4**, 562–566.
17. S. P. Surwade, S. N. Smirnov, I. V. Vlassiouk, R. R. Unocic, G. M. Veith, S. Dai and S. M. Mahurin, *Nat. Nanotechnol.*, 2015, **10**, 459–464.

18. S. C. O'Hern, D. Jang, S. Bose, J.-C. Idrobo, Y. Song, T. Laoui, J. Kong and R. Karnik, *Nano Lett.*, 2015, **15**, 3254–3260.
19. K. S. Novoselov, A. K. Geim, S. V. Morozov, D. Jiang, Y. Zhang, S. V. Dubonos, I. V. Grigorieva and A. A. Firsov, *Science*, 2004, **306**, 666–669.
20. W.-W. Liu, S.-P. Chai, A. R. Mohamed and U. Hashim, *Synthesis and Characterization of Graphene and Carbon Nanotubes: A Review on the Past and Recent Developments*, 2014, vol. 20.
21. S. Bhaviripudi, X. Jia, M. S. Dresselhaus and J. Kong, *Nano Lett.*, 2010, **10**, 4128–4133.
22. X. Li, W. Cai, J. An, S. Kim, J. Nah, D. Yang, R. Piner, A. Velamakanni, I. Jung, E. Tutuc, S. K. Banerjee, L. Colombo and R. S. Ruoff, *Science*, 2009, **324**, 1312–1314.
23. F. M. Kafiah, Z. Khan, A. Ibrahim, R. Karnik, M. Atieh and T. Laoui, *Desalination*, 2016, **388**, 29–37.
24. M. P. Lavin-Lopez, A. Paton-Carrero, L. Sanchez-Silva, J. L. Valverde and A. Romero, *Adv. Powder Technol.*, 2017, **28**, 3195–3203.
25. S. Dubin, S. Gilje, K. Wang, V. C. Tung, K. Cha, A. S. Hall, J. Farrar, R. Varshneya, Y. Yang and R. B. Kaner, *ACS Nano*, 2010, **4**, 3845–3852.
26. L. Qiu, X. Zhang, W. Yang, Y. Wang, G. P. Simon and D. Li, *Chem. Commun. (Cambridge, U. K.)*, 2011, **47**, 5810–5812.
27. A. Akbari, P. Sheath, S. T. Martin, D. B. Shinde, M. Shaibani, P. C. Banerjee, R. Tkacz, D. Bhattacharyya and M. Majumder, *Nat. Commun.*, 2016, 7, 10891.
28. K. A. Mahmoud, B. Mansoor, A. Mansour and M. Khraisheh, *Desalination*, 2015, **356**, 208–225.
29. A. Giwa, N. Akther, V. Dufour and S. W. Hasan, *RSC Adv.*, 2016, **6**, 8134–8163.
30. S. Daer, J. Kharraz, A. Giwa and S. W. Hasan, *Desalination*, 2015, **367**, 37–48.
31. H. M. Hegab and L. Zou, *J. Membr. Sci.*, 2015, **484**, 95–106.
32. L. He, L. F. Dumée, C. Feng, L. Velleman, R. Reis, F. She, W. Gao and L. Kong, *Desalination*, 2015, **365**, 126–135.
33. H. Chae, J. Lee, C. Lee, I. Kim and P. Park, *J. Memb. Sci.*, 2015, **483**, 128–135.
34. W. Choi, J. Choi, J. Bang and J. Lee, *ACS Appl. Mater. Interfaces*, 2013, **5**, 12510–12519.
35. F. Shao, L. Dong, H. Dong, Q. Zhang, M. Zhao, L. Yu, B. Pang and Y. Chen, *J. Membr. Sci.*, 2017, **525**, 9–17.
36. M. Safarpour, A. Khataee and V. Vatanpour, *J. Membr. Sci.*, 2015, **489**, 43–54.
37. K. Chen, C. Xiao, Q. Huang, H. Liu and Y. Tang, *Desalination*, 2017, **425**, 175–184.
38. H. J. Kim, Y.-S. Choi, M.-Y. Lim, K. H. Jung, D.-G. Kim, J.-J. Kim, H. Kang and J.-C. Lee, *J. Membr. Sci.*, 2016, **514**, 25–34.
39. M. Fathizadeh, H. N. Tien, K. Khivantsev, Z. Song, F. Zhou and M. Yu, *Desalination*, 2017, 1–8.

40. H. Salehi, M. Rastgar and A. Shakeri, *Appl. Surf. Sci.*, 2017, **413**, 99–108.
41. H. M. Hegab, A. ElMekawy, T. G. Barclay, A. Michelmore, L. Zou, C. P. Saint and M. Ginic-Markovic, *Desalination*, 2016, **385**, 126–137.
42. X. Wang, X. Wang, P. Xiao, J. Li, E. Tian, Y. Zhao and Y. Ren, *Colloids Surfaces A Physicochem. Eng. Asp.*, 2016, **508**, 327–335.
43. M. Bhadra, S. Roy and S. Mitra, *Desalination*, 2016, **378**, 37–43.
44. F. U. Nigiz, *Desalination*, 2017, **433**, 164–171.
45. K. Xu, B. Feng, C. Zhou and A. Huang, *Chem. Eng. Sci.*, 2016, **146**, 159–165.
46. B. Feng, K. Xu and A. Huang, *Desalination*, 2016, **394**, 123–130.
47. D. Cohen-Tanugi, L. C. Lin and J. C. Grossman, *Nano Lett.*, 2016, **16**, 1027–1033.
48. D. Cohen-Tanugi and J. C. Grossman, *J. Chem. Phys.*, 2014, **141**, 74704.
49. D. H. Cohen-tanugi, *Nano Lett.*, 2012, **12**, 3602–3608.
50. J.-G. Gai and X.-L. Gong, *J. Mater. Chem. A*, 2014, **2**, 425–429.
51. Y. Li, Z. Xu, S. Liu, J. Zhang and X. Yang, *Comput. Mater. Sci.*, 2017, **139**, 65–74.
52. G. Wu, X. Tang, M. Meyyappan and K. W. C. Lai, *Appl. Surf. Sci.*, 2017, **425**, 713–721.
53. Q. Chen and X. Yang, *J. Membr. Sci.*, 2015, **496**, 108–117.
54. V. Singh, D. Joung, L. Zhai, S. Das, S. I. Khondaker and S. Seal, *Prog. Mater. Sci.*, 2011, **56**, 1178–1271.
55. X. Li, C. W. Magnuson, A. Venugopal, R. M. Tromp, J. B. Hannon, E. M. Vogel, L. Colombo and R. S. Ruoff, *J. Am. Chem. Soc.*, 2011, **133**, 2816–2819.
56. A. Zurutuza and C. Marinelli, *Nat. Nanotechnol.*, 2014, **9**, 730–734.

CHAPTER 10

Graphene and 2D Materials Based Membranes for Water Treatment

ANNAROSA GUGLIUZZA,*[a] ANTONIO POLITANO[b] AND ENRICO DRIOLI*[a,c]

[a] Research Institute on Membrane Technology-National Research Council (CNR-ITM), Via Pietro Bucci 17C, Rende (CS), 87036, Italy; [b] Fondazione Istituto Italiano di Tecnologia, Graphene Labs, Via Morego 30, 16163 Genoa, Italy; [c] Department of Energy Engineering, College of Engineering, Hanyang University, Seoul 133-791, Republic of Korea
*Email: a.gugliuzza@itm.cnr.it; e.drioli@itm.cnr.it

10.1 Introduction

Two-dimensional materials (2D) are the thinnest functional nanomaterials, due to their atomic scale thickness and infinite planar dimensions. They are characterized by strong in-plane chemical bonds and weak interlayer bonds.[1] By means of micromechanical[2] or chemical[3] exfoliation techniques, it is possible to isolate a few or single freestanding nanosheets, which can be used as atomically thick membranes or, alternatively, as nanofillers in polymeric membranes, thus yielding novel functionalities. The existence of a large and chemically active surface area in nanoporous 2D materials inside composite membranes may significantly improve their transport properties.

The recent achievements of membrane technology in different fields, from water desalination[4,5] to gas separation,[6,7] are based on research

Nanoscience & Nanotechnology Series No. 47
Graphene-based Membranes for Mass Transport Applications
Edited by Pengzhan Sun and Hongwei Zhu

Published by the Royal Society of Chemistry, www.rsc.org

on (i) materials and (ii) optimization of processes. By adopting standard solutions, further noteworthy advancement is unfeasible. Conversely, unexplored materials could provide interesting perspectives. As an example, the use of 2D materials could afford new inputs to the production of nanoporous desalination membranes, through which water flow and ion filtering can be regulated. Also, liquid and gas membrane operations, different from the traditional and already dominant pressure-driven membrane separation systems, might benefit from the employment of 2D materials.

Graphene, the forefather of 2D materials,[8] is entirely impermeable to most molecular compounds,[9] due to the localization of electrons within aromatic rings, which form a faultless barrier to the passage of molecules. Nevertheless, the unimpeded permeation of water through graphene-based membranes[10,11] has revealed the existence of unique conditions for molecular dynamics in 2D spaces.[12] However, the presence of defects in atomically thick graphene membranes is unavoidable with subsequent implications for membrane separations.[6]

Further computational and experimental efforts are still necessary to identify reliable and feasible routes for the large-scale availability of graphene-based membranes, in which the interplay of morphological and chemical characteristics yields outstanding separation performances.

Following the extensive investigation of graphene in the last decade,[5,6,13,14] other 2D materials "beyond graphene"[15,16] are being considered for advanced applications in membrane technology.[17,18] The synthesis of novel 2D materials,[16] such as transition-metal dichalcogenides (TMDCs), 2D carbides, IV–VI compounds or atomically thin elemental materials – silicene, germanene, phosphorene, and borophene – is expected to open new pathways, since these materials show exotic physicochemical properties, which have never been accessed so far with three-dimensional (3D) materials.[16]

10.2 Membrane Processes for Water Desalination: A Brief Outlook

Membrane science and technology has been extensively developed in recent years for a broad range of applications, including water desalination.[19] Some membrane operations from reverse osmosis (RO) to electrodialysis (ED) and membrane distillation (MD) are mainly dedicated to water desalination and share the need to have nanostructured membranes designed for enhanced productivity–efficiency trade-offs.[20,21] Among the pressure-driven processes, RO is one of the most exploited operations for water desalination. This process uses membranes working as a semipermeable interface between two solutions at different concentrations. The water transfer through the membrane is proportional to the pressure differential that exceeds the natural osmotic pressure differential. High operating pressures are required for the initial pressurization of the feed, *i.e.* 15–30 bar for brackish water desalination and 55–70 bar for seawater desalination.

ED has been proposed for small- and medium-scale desalination of brackish water and consists of applying an electrical field to remove salt ions from water using a pair of cation and anion exchange membranes, respectively. Specifically, the cations migrate from the brackish water toward the negative electrode through the cation exchange membrane, while anions migrate toward the anode through the anion exchange membrane.

MD is part of the membrane contactor processes, which induce mass transfer between two phases without dispersing one phase into the other.

In Direct Contact Membrane Distillation (DCMD), hydrophobic microporous membranes are used to separate two streams of water, preventing liquid intrusion inside the pores while water vapor passes freely to be condensed on the permeate side. The driving-force of the process is usually a water vapor pressure difference, which can be induced by a difference of temperature (thermally driven MD) or a difference of solute concentration (osmotic MD). In thermally driven MD, the hydrophobic membrane separates a hot and cold stream of water. In osmotic MD, the water vapor transfer is the effect of the concentration of feed and the dilution of the osmotic agent solution.[22] Other MD configurations include Air Gap Membrane Distillation (AGMD), Sweeping Gas Membrane Distillation (SGMD), and Vacuum Membrane Distillation (VMD).

Concerning DCMD modes, both techniques have the advantage over traditional RO of working at much lower pressure and temperature, resulting in competitive energy consumption. In addition, a higher rejection factor and lower susceptibility to concentration polarization phenomena are achieved. According to this approach, membrane crystallization (MCr) represents a further advanced application, enabling the retrieval of salts with uniform shape and size distribution, providing desalted water.[23,24] Another category is represented by membrane condenser (MC) for selective recovery of water from the atmosphere and, in particular, from the exhaust gases produced in many industrial production processes as a further new source of drinkable water.[25] In all these cases, the major concern is related to the unavailability of membranes with structural and chemical features suitable for selective passage of vapor without contamination, scaling and polarization effects. A common foremost ambition is to get ultrathin interfaces in which short and non-tortuous pathways can be combined with durable wetting resistance and, where required, controlled heat loss. The combination of all these elements is expected to provide the outstanding productivity–efficiency trade-off necessary for scalability at industrial levels. 2D materials seem to have suitable characteristics to obtain the desired targets, which are necessary to overcome the current limitations of membrane separation processes dedicated to water treatment.[26]

10.3 Porous Graphene as a Precursor for Ultrathin Membranes

The potential use of graphene as a membrane implicates the drilling of the nanosheet in order to generate sub-nanometer pores through which the smallest molecules can pass, whereas the biggest ones are blocked. An atomically thick membrane section and nanosized pores are envisaged to enhance the productivity–efficiency trade-off for membrane desalination processes. High water flux and outstanding rejection may have positive repercussions on the overall efficiency of the separation process.

However, the major concern is the identification of a practical and reliable route to produce nanopores in graphene sheets in a controlled and scalable way. Various approaches, including (i) high-energy electron and ion bombardment,[27] (ii) ultraviolet-induced oxidative[28] and oxygen-plasma etching,[5] as well as molecular building blocks combined with nano-imprinting lithography,[29] have been proposed to produce sub-nanopores in graphene sheets.[30] However, whenever sub-nanometer pores are formed, some troublesome events can take place, including shrinkage and expansion of graphene nanopores and self-healing. Specifically, when using an electron beam at temperatures in the 400–1200 °C range, pores of 10 nm can be completely closed. A possible solution is to stop the electron beam process at a suitable time.[31] Also, interactions with metal impurities could catalyze undesired self-healing.[32] Recently, Robertson *et al.*[33] presented the first image with atomic resolution of sub-nanometer pores in graphene by keeping graphene at ~500 °C in order to prevent self-healing of nanopores with mobile carbon species. This approach based on transmission electron microscopy (TEM) remains, however, too expensive for mass-production. As an alternative, Zhang *et al.*[34] used scanning electron microscopy (SEM) with the assistance of nitrogen ions and an electron beam of less than 10 keV to generate nanopore patterns in graphene layers.

An electrical pulse fabrication method based on a simple fluidic cell has been proposed for the production of single nanopores down to sub-nanometer sizes. Short voltage pulses were applied across the membrane in electrolyte solution to create or enlarge nanopores in order to remove the minimum amount of material and achieve a higher control of pore size.[35]

Induced oxidative etching approaches, commonly used for preparing polymeric membranes, together with bombardment with gallium ions, have been proposed for creating pores in graphene sheets.[36] Bunch and coworkers[28] exposed graphene, after mechanical exfoliation and pressurization with H_2, to ultraviolet light under ambient conditions to create nanopores.

Block copolymers have been proposed to form an annealed layer on graphene sheets and, then, this was covered with an *in situ* formed SiO_x-based nanopore hard mask. Next, O_2-plasma etching was used to drill the graphene and obtain regular pores through the sheet once the SiO_x mask was lifted.[37,38]

Chemical activation by KOH and combined thermal shock or microwave irradiation were also employed to produce GO with a specific surface area (SSA) up to 1900 $m^2\,g^{-1}$ and pore volume up to 1.65 $cm^3\,g^{-1}$.[39]

Layer-by-layer (LbL) assembly has also been used to create highly tunable thin and functional films with nanometer-scale control over the film composition and structure.[40,41] Charged chemical groups have been introduced over a wide range of pH conditions, thus producing chemically functionalized exfoliated GO sheets.[42]

The major concern that remains, however, is the control of pore size and pore distribution at the atomic scale, thus producing serious restrictions, due to the risk of selectivity loss. In this regard, thermal activation of aerogel graphene has been used to introduce pores in graphene sheets in a controlled and scalable way.[43] The reduction of GO and the exploitation of different thermal resistances of two amorphous and disordered domains and highly crystalline domains have been suggested as a practical route for the mass-production of nanoporous graphene with uniform pore size.

For membrane applications, the attractive challenge is to attain scalable production of nanosized and uniform pores in ultrathin layers in order to provide high a permeability–selectivity trade-off, reduced energy consumption together with controlled (bio)fouling. 2D materials represent an interesting opportunity to realize atomically thick membranes with peculiar physicochemical properties, which could allow overcoming the current limits for industrial-scale membrane separation operations.

10.4 Graphene and Water Desalination

Ion filtering represents one of the most promising applications of graphene membranes, especially *via* RO and ED operations.[5,29,44–47] Some studies based on molecular dynamics simulations have shown that water transport through a porous graphene membrane is more efficient in comparison with the case of carbon nanotubes and other polymeric membranes[4,27] (Figure 10.1), suggesting a salt rejection–water permeability trade-off around 100 for hydrogenated and hydroxylated graphene. This target is justified by the atomic thickness, which requires less pressure and, hence, less energy to push water through it.

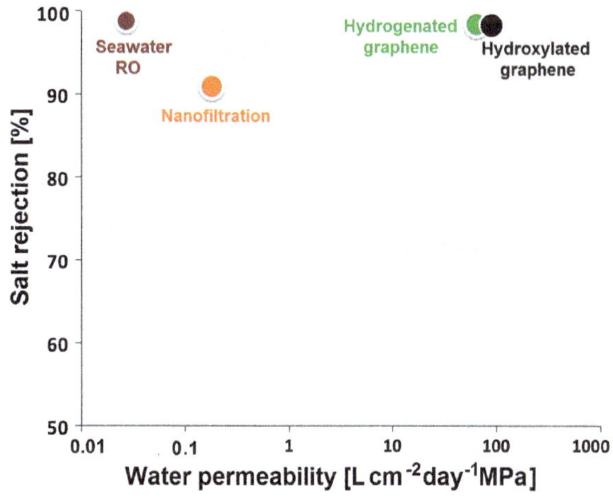

Figure 10.1 Salt rejection *vs.* water permeability for standard polymeric and graphene-based membranes.

Pores in graphene achieved by short oxidation times have shown a cation-selective behavior, due to electrostatic repulsions caused by negatively charged functional groups terminating the pore edges.

On the contrary, a longer oxidation has been demonstrated to cause steric size exclusion. Monovalent cations such as K^+ over Cl^- have been removed with a selectivity ratio over 100 through voltage-activated channels,[48] while stacked graphene oxide layer membranes have been demonstrated to retain hydrated salt ions and let water molecules pass through.[49]

The length of nanochannels in graphene has been envisaged to be a key parameter to remove ions from water by applying a concentration gradient-driven diffusion rather than high pressure.[50]

Clearly, the rearrangement of materials determines the selective pathway and interactions ruling ion and liquid transport.[51–58] As an example, Figure 10.2 shows a comparison of water flux and NaCl rejection values experimentally estimated for rGO- and GO-containing membranes. In most cases, 100% NaCl rejection is obtained, while water permeance has been found to be slightly higher than in commercial SWRO and brackish water reverse osmosis (BWRO) membranes, thus confirming the attractive potential of membranes based on 2D materials.

Indeed, a univocal identification of structure–transport relationships, enabling one to regulate a separation mechanism rather than another, does not exist. Recently, Sun *et al.*[59] investigated the possible mechanisms involving ion transport through GO membranes, examining structural and chemical effects on the diffusion of differently sized molecules through modified graphene (Figure 10.3). It was evidenced that the formation of nanocapillary interlayer spacing between GO sheets and the establishment of non-covalent π interactions enable guest molecules to be accommodated or quickly pass.[60] The presence of hydrophilic groups favors an increase in the interlayer stacking up to 1.4 nm.[61] Depending on the interlayer spacing, small solutes can be separated from species with larger hydrated radii according to size sieving mechanisms, while hydroxylated chemical moieties locally change the hybridization of carbon atoms from sp^2 to sp^3, modifying the interaction between guest molecules and the matrix. An interlayer hexagonal ice bilayer[12] was envisaged to form and melt at the sheet edges, allowing monovalent cations and strong acids or bases to diffuse more quickly than bivalent cations or weak acids and bases. However, sp^2 regions within sp^3 systems could promote selective π–π* interactions[62,63] between transition metals and functional groups attached to GO sheets, so conferring to the membrane a real ion-selectivity character. The predominance of regions with sp^2 hybridization allows the membrane to discriminate among cations without *d* electrons. A recent density functional theory (DFT) study quantified the interactions between alkali and alkaline earth cations with counter Cl^- anions and nanosized and microsized graphene membranes within a range of 10 to 40 °C,[59] confirming that cations without *d* electrons interact with the sp^2 clusters *via* cation–π interactions and transition metal ions coordinate to sp^3-configured sites of the membrane.

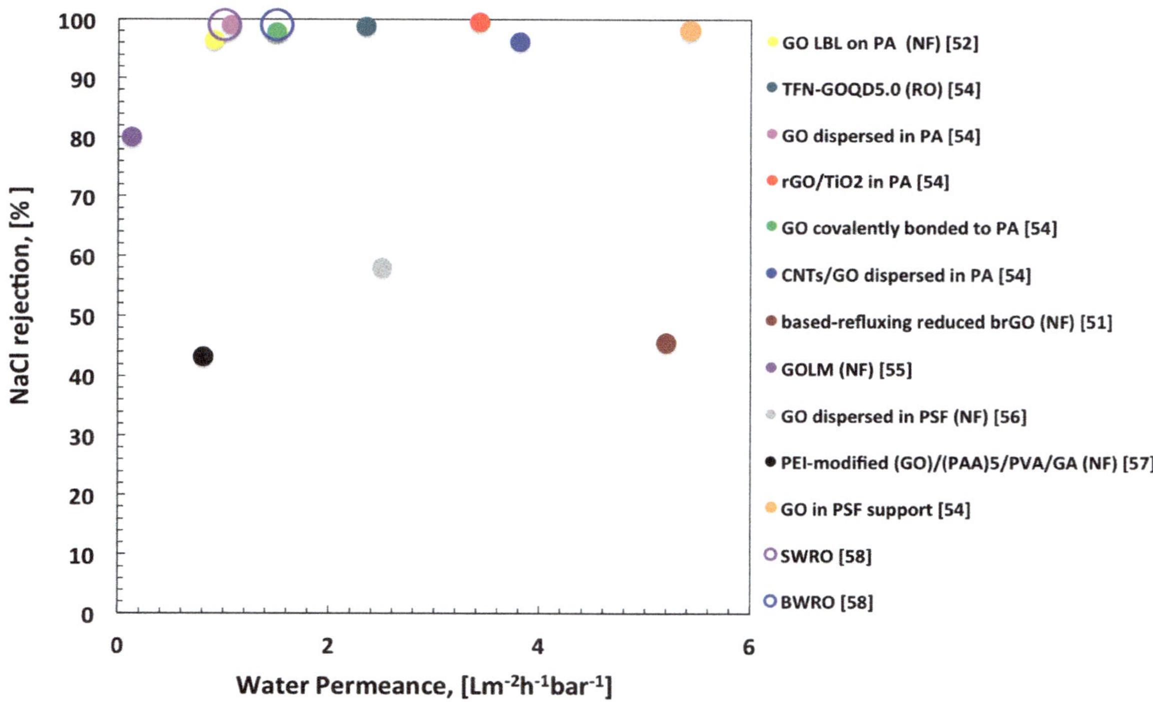

Figure 10.2 Experimental water permeance and NaCl rejection of traditional polymeric RO membranes and GO-containing RO membranes. GOQD stands for GO quantum dots, while GOLM stands for GO laminate membranes.
Reprinted from *Current Opinion in Chemical Engineering*, **16**, A. Gugliuzza, A. Politano and E. Drioli, The advent of graphene and other two-dimensional materials in membrane science and technology, 78–85, Copyright 2017, with permission from Elsevier.[26]

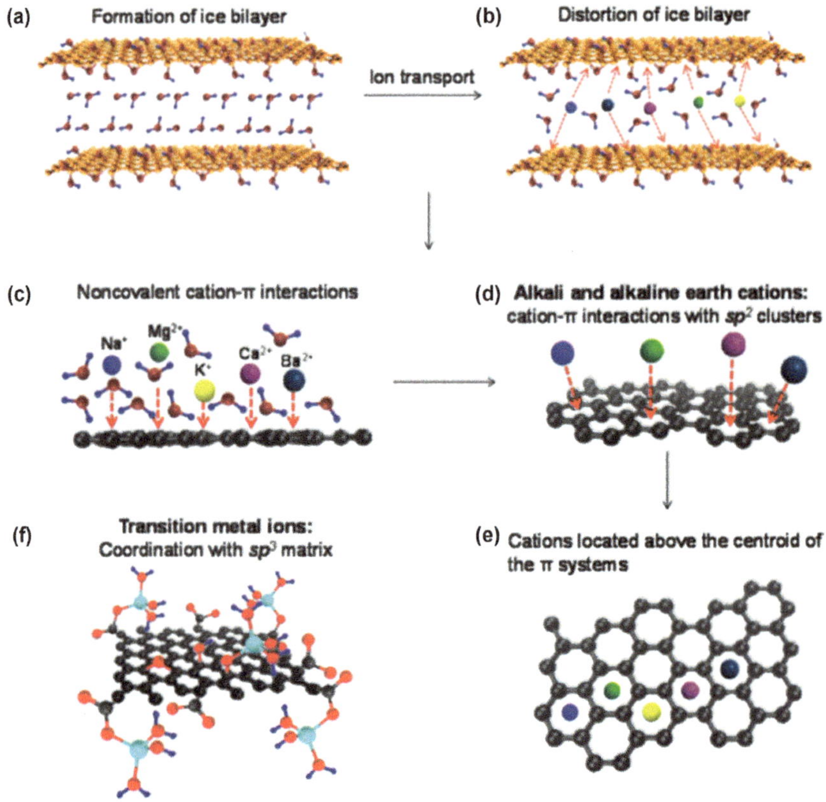

Figure 10.3 Proposed mechanisms of transport through GO laminar membranes: (a) ice bilayer within interlayered GO flakes; (b) distorted structure of ice bilayer due to water–cation interactions; (c–e) noncovalent cation–π interactions of alkali and alkaline earth cations with sp^2 clusters of GO sheet; (f) coordination of transition metal ions with sp^3 matrix of GO sheets.
Reprinted with permission from P. Sun, F. Zheng, M. Zhu, Z. Song, K. Wang, M. Zhong, D. Wu, R. B. Little, Z. Xu and H. Zhu, *ACS Nano*, 2014, **8**, 850, Copyright 2014 American Chemical Society.[59]

Actually, a strong discrepancy between theoretical and experimental studies exists. This gap can be filled only through elaborating new models and software and establishing practicable and feasible manufacturing procedures at desired scales.

10.5 Potential use of 2D Materials in Advanced Membrane Contactor Operations

MD, MCr and MC are three distinctive categories of membrane contactors, which have in common the requirement to use nanoporous hydrophobic

Table 10.1 Thermal conductivity of 2D materials, as compared with polymers used in MD. Reprinted from *Current Opinion in Chemical Engineering*, **16**, A. Gugliuzza, A. Politano and E. Drioli, The advent of graphene and other two-dimensional materials in membrane science and technology, 78–85, Copyright 2017, with permission from Elsevier.[26]

Material	Thermal conductivity @ 300 K [$W\ m^{-1} K^{-1}$]	Reference
Monolayer graphene	~5000	64
Monolayer SnSe	0.46–0.68 (anisotropic)	65
Sb_2Te_3	0.23–0.5	66
Bi_2Se_3	0.4 (few layers)	67
MoS_2	108 (monolayer)	68
WSe_2	0.05 (monolayer) 1.5 (bulk)	69
Phosphorene	zigzag direction: 110 armchair direction: 36	70
PVDF	0.19	
Polytetrafluoroethylene	0.24–0.35	
Polypropylene	0.10–0.22	
Cellulose acetate	0.16–0.36	

membranes with ultrathin thickness, low tortuosity factor, high porosity, permanent anti-wetting properties and controlled thermal conductivity. Graphene may be an interesting material due to its robustness, impermeability and atomically thick nature. However, it exhibits outstanding thermal conductivity (~5000 $Wm^{-1} K^{-1}$ at 300 K), which could be a strong limitation for thermally driven membrane operations such as MD and MCr, under specific conditions, while seeming beneficial for MC. Although thermal conductivity properties are still unexplored in relation to changes in sample size, strain or defects, other classes of atomically thick materials, such as hexagonal boron nitride (h-BN), silicene, TMDCs, transition metal oxides, five-layered V–VI trichalcogenides and black phosphorus, with competitive thermal conductivity are currently considered as a possible alternative (Table 10.1).

An explorative study has demonstrated that bismuth selenide (Bi_2Se_3) used as a filler in PVDF membranes can have effects on the crystallization events of salts. MCr experiments carried out with NaCl solutions (4 M) have, indeed, confirmed that Bi_2Se_3 preserves the excellent resistance of the membrane to wetting. Very high NaCl rejection factors have been achieved, resulting in a faster crystal growth rate.[71] Despite being preliminary, this explorative study indicates the possible attractive effects of 2D materials on nucleation and the growth rate of crystals (Figure 10.4).

10.6 2D Materials Beyond Graphene for Ion Filtering

Chemically exfoliated monolayers of molybdenum disulfide (MoS_2),[72,73] tungsten disulfide (WS_2),[74] and MXene[75] have been manipulated to form membranes.[76] One of the most studied is MoS_2 due to its cheaper and higher

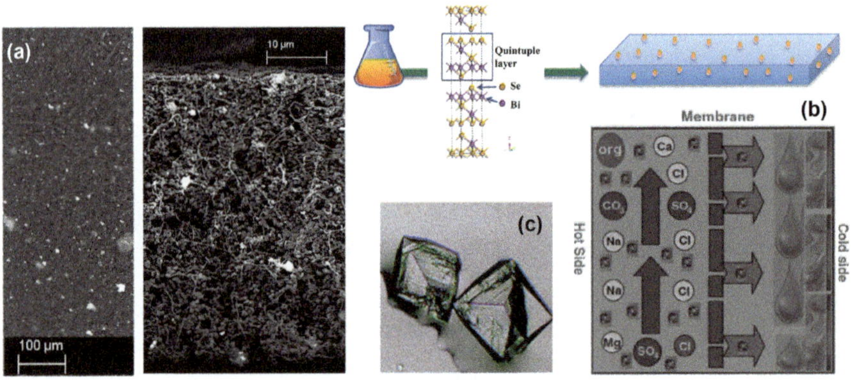

Figure 10.4 (a) SEM micrographs of nanocomposite PVDF/Bi_2Se_3 membranes; (b) scheme of membrane crystallization process; (c) salt crystals formed after working the nanocomposite membranes in membrane crystallizer.

performing behavior concerning the capability to transport water and reject salts with a retention of more than 88% of ions. A water flux 2–5 orders of magnitude higher than that of traditional nanoporous membranes and 70% greater than that of porous graphene has been modeled for this kind of membrane.[18]

In particular, water flux and ion rejection have been simulated through three pore edge types for MoS_2 (Mo only, S only, and Mo + S mixed), having an accessible area comparable to graphene (mixed, $A = 55.45$ Å^2; Mo only, $A = 56.42$ Å^2; S only, $A = 57.38$ Å^2; and graphene, $A = 59.67$ Å^2). As a result, the water flux through the mixed pore resulted in being intermediary between Mo only and S only nanopores.

Flat MoS_2 membranes have also been produced with 3 nm sized pores, showing a water flux 3 to 5 times higher than that of graphene oxide membranes and exhibiting a half-life of 1 week, due to the absence of hydrophilic moieties.[72] Tunable pores in monolayer MoS_2 have been simulated and mechanical stretching (up to 12%) has been envisaged as an 'open–closed' pore switcher, resulting in a passage of water up to 2 orders of magnitude higher than that of the commercial RO membranes.[17] It has also been found that both steric and electrostatic effects provide the membrane with a different filtering capability toward monovalent ions. Sodium is solvated by water molecules, forming a shell much tighter than that around chloride ions and the related free energy to extract bound water molecules is 9.90 kJ mol^{-1} against the value of 4.36 kJ mol^{-1} necessary for hydrated chloride anions. This implies a water exchange time for Na^+ of 27.22 ± 0.26 ps, while 8.58 ± 0.03 ps is needed for Cl^-.

Experimentally, the joint effect of steric hindrance, compatibility with the 2D hydrogen bonding network of water and electrostatic interactions has been envisaged as responsible for aqueous ion diffusion through Mo_2S lamellar membranes, while a molecular sieving property for organic vapor has been recently evidenced.[77]

It is worth noting that the lack of a 3D counterpart of silicene, germanene, and borophene samples complicates their effective utilization in membrane technology, since these single-layer elemental materials can be grown only on single-crystal surfaces by chemical vapor deposition.[78] In contrast, phosphorene can be exfoliated from bulk crystals of black phosphorus. However, phosphorene layers are extremely reactive toward oxygen and water[79] in standard conditions, even if ambient stability could be achieved.[80] The insufficient surface chemical stability of phosphorene hinders its extensive technological use as an ultrathin membrane, in contrast with the case of ambient-stable graphene. On the contrary, $Ti_3C_2T_x$ layers, which are part of the MXene family, have been proposed for the separation of monovalent cations. Due to their hydrophilic nature, water passage through the interspaces is facilitated with a flux of up to 37.4 $Lh^{-1}\,m^{-2}\,bar^{-1}$.[75]

10.7 Conclusions

Membrane processes involving 2D materials enable novel functional applications for water treatment and salt recovery. The attractive prospect to use atomically thick layers in some areas of membrane science originates from the goal to get a reduced molecular pathway, high discrimination factor, and long-term wetting resistance with, in some cases, relatively controlled thermal conductivity.

Preliminary studies are providing encouraging and promising evidence for the tremendous opportunity offered by 2D materials for advanced wastewater treatment and, in particular, ion filtering. The design of nanofilters should benefit from the stability of inorganic nanomaterials based on strain control. Nevertheless, massive production of 2D materials is still unaffordable compared with state-of-the-art polymeric membranes, while comparative experimental and theoretical studies are still necessary to identify suitable structure–transport relationships for designing scalable and affordable novel membranes with outstanding separation performance.

References

1. A. Gupta, T. Sakthivel and S. Seal, *Prog. Mater. Sci.*, 2015, **73**, 44.
2. B. Jakob, *Europhys. Lett.*, 2015, **109**, 58004.
3. J. N. Coleman, M. Lotya, A. O'Neill, S. D. Bergin, P. J. King, U. Khan, K. Young, A. Gaucher, S. De, R. J. Smith, I. V. Shvets, S. K. Arora, G. Stanton, H.-Y. Kim, K. Lee, G. T. Kim, G. S. Duesberg, T. Hallam, J. J. Boland, J. J. Wang, J. F. Donegan, J. C. Grunlan, G. Moriarty, A. Shmeliov, R. J. Nicholls, J. M. Perkins, E. M. Grieveson, K. Theuwissen, D. W. McComb, P. D. Nellist and V. Nicolosi, *Science*, 2011, **331**, 568.
4. D. Cohen-Tanugi and J. C. Grossman, *Nano Lett.*, 2012, **12**, 3602.
5. S. P. Surwade, S. N. Smirnov, I. V. Vlassiouk, R. R. Unocic, G. M. Veith, S. Dai and S. M. Mahurin, *Nat. Nanotechnol.*, 2015, **10**, 459.

6. M. S. H. Boutilier, C. Sun, S. C. O'Hern, H. Au, N. G. Hadjiconstantinou and R. Karnik, *ACS Nano*, 2014, **8**, 841.
7. M. Shan, Q. Xue, N. Jing, C. Ling, T. Zhang, Z. Yan and J. Zheng, *Nanoscale*, 2012, **4**, 5477.
8. K. S. Novoselov, *Rev. Mod. Phys.*, 2011, **83**, 837.
9. V. Berry, *Carbon*, 2013, **62**, 1.
10. R. R. Nair, H. A. Wu, P. N. Jayaram, I. V. Grigorieva and A. K. Geim, *Science*, 2012, **335**, 442.
11. Y. Su, V. Kravets, S. Wong, J. Waters, A. Geim and R. Nair, *Nat. Commun.*, 2014, **5**, 4843.
12. D. W. Boukhvalov, M. I. Katsnelson and Y.-W. Son, *Nano Lett.*, 2013, **13**, 3930.
13. A. C. Ferrari, F. Bonaccorso, V. Fal'ko, K. S. Novoselov, S. Roche, P. Boggild, S. Borini, F. H. L. Koppens, V. Palermo, N. Pugno, J. A. Garrido, R. Sordan, A. Bianco, L. Ballerini, M. Prato, E. Lidorikis, J. Kivioja, C. Marinelli, T. Ryhanen, A. Morpurgo, J. N. Coleman, V. Nicolosi, L. Colombo, A. Fert, M. Garcia-Hernandez, A. Bachtold, G. F. Schneider, F. Guinea, C. Dekker, M. Barbone, Z. Sun, C. Galiotis, A. N. Grigorenko, G. Konstantatos, A. Kis, M. Katsnelson, L. Vandersypen, A. Loiseau, V. Morandi, D. Neumaier, E. Treossi, V. Pellegrini, M. Polini, A. Tredicucci, G. M. Williams, B. Hee Hong, J.-H. Ahn, J. Min Kim, H. Zirath, B. J. van Wees, H. van der Zant, L. Occhipinti, A. Di Matteo, I. A. Kinloch, T. Seyller, E. Quesnel, X. Feng, K. Teo, N. Rupesinghe, P. Hakonen, S. R. T. Neil, Q. Tannock, T. Lofwander and J. Kinaret, *Nanoscale*, 2015, **7**, 4598.
14. M. E. Suk and N. R. Aluru, *J. Phys. Chem. Lett.*, 2010, **1**, 1590.
15. G. R. Bhimanapati, Z. Lin, V. Meunier, Y. Jung, J. Cha, S. Das, D. Xiao, Y. Son, M. S. Strano, V. R. Cooper, L. Liang, S. G. Louie, E. Ringe, W. Zhou, S. S. Kim, R. R. Naik, B. G. Sumpter, H. Terrones, F. Xia, Y. Wang, J. Zhu, D. Akinwande, N. Alem, J. A. Schuller, R. E. Schaak, M. Terrones and J. A. Robinson, *ACS Nano*, 2015, **9**, 11509.
16. P. Miro, M. Audiffred and T. Heine, *Chem. Soc. Rev.*, 2014, **43**, 6537.
17. W. Li, Y. Yang, J. K. Weber, G. Zhang and R. Zhou, *ACS Nano*, 2016, **10**, 1829.
18. M. Heiranian, A. B. Farimani and N. R. Aluru, *Nat. Commun.*, 2015, **6**, 8616.
19. M. Elimelech and W. A. Phillip, *Science*, 2011, **333**, 712.
20. V. Speranza, F. Trotta, E. Drioli and A. Gugliuzza, *ACS Appl. Mater. Interfaces*, 2010, **2**, 459.
21. M. Perrotta, G. Saielli, G. Casella, F. Macedonio, L. Giorno, E. Drioli and A. Gugliuzza, *Appl. Mater. Today*, 2017, **9**, 1.
22. E. Drioli, A. Ali and F. Macedonio, *Desalination*, 2015, **356**, 56.
23. E. Drioli, G. Di Profio and E. Curcio, *Curr. Opin. Chem. Eng.*, 2012, **1**, 178.
24. A. Gugliuzza, M. C. Aceto and E. Drioli, *Polym. Int.*, 2009, **58**, 1452.
25. F. Macedonio, A. Brunetti, G. Barbieri and E. Drioli, *Ind. Eng. Chem. Res.*, 2013, **52**, 1160.

26. A. Gugliuzza, A. Politano and E. Drioli, *Curr. Opin. Chem. Eng.*, 2017, **16**, 78.
27. K. Celebi, J. Buchheim, R. M. Wyss, A. Droudian, P. Gasser, I. Shorubalko, J.-I. Kye, C. Lee and H. G. Park, *Science*, 2014, **344**, 289.
28. S. P. Koenig, L. Wang, J. Pellegrino and J. S. Bunch, *Nat. Nanotechnol.*, 2012, 7, 728.
29. M. Hu and B. Mi, *J. Membr. Sci.*, 2014, **469**, 80.
30. W. Yuan, J. Chen and G. Shi, *Mater. Today*, 2014, **17**, 77.
31. N. Lu, J. Wang, H. C. Floresca and M. J. Kim, *Carbon*, 2012, **50**, 2961.
32. R. Zan, Q. M. Ramasse, U. Bangert and K. S. Novoselov, *Nano Lett.*, 2012, **12**, 3936.
33. A. W. Robertson, G.-D. Lee, K. He, C. Gong, Q. Chen, E. Yoon, A. I. Kirkland and J. H. Warner, *ACS Nano*, 2015, **9**, 11599.
34. D. Fox, A. O'Neill, D. Zhou, M. Boese, J. Coleman and H. Zhang, *Appl. Phys. Lett.*, 2011, **98**, 243117.
35. A. T. Kuan, B. Lu, P. Xie, T. Szalay and J. A. Golovchenko, *Appl. Phys. Lett.*, 2015, **106**, 203109.
36. S. C. O'Hern, M. S. Boutilier, J.-C. Idrobo, Y. Song, J. Kong, T. Laoui, M. Atieh and R. Karnik, *Nano Lett.*, 2014, **14**, 1234.
37. J. Bai, X. Zhong, S. Jiang, Y. Huang and X. Duan, *Nat. Nanotechnol.*, 2010, **5**, 190.
38. B. H. Kim, J. Y. Kim, S.-J. Jeong, J. O. Hwang, D. H. Lee, D. O. Shin, S.-Y. Choi and S. O. Kim, *ACS Nano*, 2010, **4**, 5464.
39. G. Srinivas, J. Burress and T. Yildirim, *Energy Environ. Sci.*, 2012, **5**, 6453.
40. V. Pingitore, D. Miriello, E. Drioli and A. Gugliuzza, *Soft Matter*, 2015, **11**, 4461.
41. A. Gugliuzza, V. Pingitore, D. Miriello and E. Drioli, *Phys. Chem. Chem. Phys.*, 2015, **17**, 12919.
42. T. Lee, S. H. Min, M. Gu, Y. K. Jung, W. Lee, J. U. Lee, D. G. Seong and B.-S. Kim, *Chem. Mater.*, 2015, **27**, 3785.
43. U. N. Maiti, R. Thapa, J. Lim, D. J. Li, K. H. Kim and S. O. Kim, *ACS Appl. Mater. Interfaces*, 2015, **7**, 25898.
44. S. C. O'Hern, D. Jang, S. Bose, J.-C. Idrobo, Y. Song, T. Laoui, J. Kong and R. Karnik, *Nano Lett.*, 2015, **15**, 3254.
45. S. Zhao, J. Xue and W. Kang, *J. Chem. Phys.*, 2013, **139**, 114702.
46. Z. He, J. Zhou, X. Lu and B. Corry, *ACS Nano*, 2013, **7**, 10148.
47. K. Sint, B. Wang and P. Král, *J. Am. Chem. Soc.*, 2008, **130**, 16448.
48. R. C. Rollings, A. T. Kuan and J. A. Golovchenko, *Nat. Commun.*, 2016, **7**, 11408.
49. Y. You, V. Sahajwalla, M. Yoshimura and R. K. Joshi, *Nanoscale*, 2016, **8**, 117.
50. P. Sun, R. Ma, H. Deng, Z. Song, Z. Zhen, K. Wang, T. Sasaki, Z. Xu and H. Zhu, *Chem. Sci.*, 2016, **7**, 6988.
51. Y. Han, Z. Xu and C. Gao, *Adv. Funct. Mater.*, 2013, **23**, 3693.
52. W. Choi, J. Choi, J. Bang and J.-H. Lee, *ACS Appl. Mater. Interfaces*, 2013, **5**, 12510.

53. I. Akin, E. Zor, H. Bingol and M. Ersoz, *J. Phys. Chem. B*, 2014, **118**, 5707.
54. X. Song, Q. Zhou, T. Zhang, H. Xu and Z. Wang, *J. Mater. Chem. A*, 2016, **4**, 16896.
55. X.-L. Xu, F.-W. Lin, Y. Du, X. Zhang, J. Wu and Z.-K. Xu, *ACS Appl. Mater. Interfaces*, 2016, **8**, 12588.
56. B. Ganesh, A. M. Isloor and A. F. Ismail, *Desalination*, 2013, **313**, 199.
57. N. Wang, S. Ji, G. Zhang, J. Li and L. Wang, *Chem. Eng. J.*, 2012, **213**, 318.
58. D. Cohen-Tanugi, R. K. McGovern, S. H. Dave, J. H. Lienhard and J. C. Grossman, *Energy Environ. Sci.*, 2014, **7**, 1134.
59. P. Sun, F. Zheng, M. Zhu, Z. Song, K. Wang, M. Zhong, D. Wu, R. B. Little, Z. Xu and H. Zhu, *ACS Nano*, 2014, **8**, 850.
60. P. Sun, M. Zhu, K. Wang, M. Zhong, J. Wei, D. Wu, Z. Xu and H. Zhu, *ACS Nano*, 2012, **7**, 428.
61. K. Raidongia and J. Huang, *J. Am. Chem. Soc.*, 2012, **134**, 16528.
62. A. S. Mahadevi and G. N. Sastry, *Chem. Rev.*, 2012, **113**, 2100.
63. J. C. Ma and D. A. Dougherty, *Chem. Rev.*, 1997, **97**, 1303.
64. A. A. Balandin, S. Ghosh, W. Bao, I. Calizo, D. Teweldebrhan, F. Miao and C. N. Lau, *Nano Lett.*, 2008, **8**, 902.
65. L.-D. Zhao, S.-H. Lo, Y. Zhang, H. Sun, G. Tan, C. Uher, C. Wolverton, V. P. Dravid and M. G. Kanatzidis, *Nature*, 2014, **508**, 373.
66. J. Chen, T. Sun, D. Sim, H. Peng, H. Wang, S. Fan, H. H. Hng, J. Ma, F. Y. C. Boey, S. Li, M. K. Samani, G. C. K. Chen, X. Chen, T. Wu and Q. Yan, *Chem. Mater.*, 2010, **22**, 3086.
67. M. K. Jana, K. Biswas and C. N. R. Rao, *Chem. – Eur. J.*, 2013, **19**, 9110.
68. W. Li, J. Carrete and N. Mingo, *Appl. Phys. Lett.*, 2013, **103**, 253103.
69. C. Chiritescu, D. G. Cahill, N. Nguyen, D. Johnson, A. Bodapati, P. Keblinski and P. Zschack, *Science*, 2007, **315**, 351.
70. A. Jain and A. J. H. McGaughey, *Sci. Rep.*, 2015, **5**, 8501.
71. E. Drioli, A. Gugliuzza and A. Politano, *Composite Graphene and Beyond Graphene Membranes, Proceeding of GraphChina 2016*, Qingdao, China, 2016.
72. L. Sun, H. Huang and X. Peng, *Chem. Commun.*, 2013, **49**, 10718.
73. D. Wang, Z. Wang, L. Wang, L. Hu and J. Jin, *Nanoscale*, 2015, 7, 17649.
74. L. Sun, Y. Ying, H. Huang, Z. Song, Y. Mao, Z. Xu and X. Peng, *ACS Nano*, 2014, **8**, 6304.
75. C. E. Ren, K. B. Hatzell, M. Alhabeb, Z. Ling, K. A. Mahmoud and Y. Gogotsi, *J. Phys. Chem. Lett.*, 2015, **6**, 4026.
76. G. Liu, W. Jin and N. Xu, *Angew. Chem.*, 2016, **55**, 13384.
77. M. Deng, K. Kwac, M. Li, Y. Jung and H. G. Park, *Nano Lett.*, 2017, **17**, 2342.
78. M. Houssa, A. Dimoulas and A. Molle, *J. Phys.: Condens. Matter*, 2015, **27**, 253002.
79. J. O. Island, G. A. Steele, H. S. J. van der Zant and A. Castellanos-Gomez, *2D Mater.*, 2015, **2**, 011002.
80. J. Pei, X. Gai, J. Yang, X. Wang, Z. Yu, D.-Y. Choi, B. Luther-Davies and Y. Lu, *Nat. Commun.*, 2016, 7, 10450.

Subject Index